Robotics in STEM Education

Myint Swe Khine
Editor

Robotics in STEM Education

Redesigning the Learning Experience

 Springer

Editor
Myint Swe Khine
Emirates College for Advanced Education
Abu Dhabi
United Arab Emirates

and

Curtin University
Bentley
Australia

ISBN 978-3-319-86249-1   ISBN 978-3-319-57786-9   (eBook)
DOI 10.1007/978-3-319-57786-9

Printed on acid-free paper

This Springer imprint is published by Springer Nature
The registered company is Springer International Publishing AG
The registered company address is: Gewerbestrasse 11, 6330 Cham, Switzerland

Contents

Part I Robotics Curriculum and Schools

1   Bringing Robotics in Classrooms   3
Amy Eguchi

**2   Systems Thinking Approach to Robotics Curriculum
in Schools**   33
Christina Chalmers and Rod Nason

**3   Combatting the War Against Machines: An Innovative
Hands-on Approach to Coding**   59
Jacqui Chetty

4   The Open Academic Robot Kit   85
Raymond K. Sheh, Amy Eguchi, Haldun Komsuoglu
and Adam Jacoff

Part II Robotics and STEM Education

5   How Have Robots Supported STEM Teaching? 103
Fabiane Barreto Vavassori Benitti and Newton Spolaôr

**6   Robotics Festival and Competitions Designed for
STEM+C Education** 131
ChanJin Chung, Christopher Cartwright and Joe DeRose

**7   Meeting Twenty-first Century Robotics and Automation
Workforce Needs in the USA** 171
Aleksandr Sergeyev

8   STEM Education by Exploring Robotics 195
Francis Tuluri

Part III   Robotics, Creativity and STEAM Education

9   **The Creative Nature of Robotics Activity:
     Design and Problem Solving**............................. 213
     Florence R. Sullivan

10 **Dancing, Drawing, and Dramatic Robots: Integrating
     Robotics and the Arts to Teach Foundational STEAM
     Concepts to Young Children** 231
     Amanda Sullivan, Amanda Strawhacker and Marina Umaschi Bers

Index 261

Part I
Robotics Curriculum and Schools

Chapter 1
Bringing Robotics in Classrooms

Amy Eguchi

Abstract Learning with educational robotics provides students, who usually are the consumers of technology, with opportunities to stop, question, and think deeply about technology. When designing, constructing, programming, and documenting the development of autonomous robots or robotics projects, students not only learn how technology works, but they also apply the skills and content knowledge learned in school in a meaningful and exciting way. Educational robotics is rich with opportunities to integrate not only STEM but also many other disciplines, including literacy, social studies, dance, music, and art, while giving students the opportunity to find ways to work together to foster collaboration skills, express themselves using the technological tool, problem-solve, and think critically and innovatively. Educational robotics is a learning tool that enhances students' learning experience through *hands-on mind-on* learning. Most importantly, educational robotics provides a fun and exciting learning environment because of its hands-on nature and the integration of technology. The engaging learning environment motivates students to learn whatever skills and knowledge needed for them to accomplish their goals in order to complete the projects of their interest. For school-age children, most robotics activities have mainly been part of informal education, such as after school programs and summer camps (Benitti in Computers & Education, 58:978–988, 2012; Eguchi 2007b; Sklar and Eguchi in Proceedings of RoboCup-2004: Robot Soccer World Cup VIII, 2004), even though it has the potential to make learning more effective in formal education. It is very difficult for teachers to include robotics in regular curriculum because of the heavy focus on standardized testing and pressure to cover academic standards set by the government and/or their States. This chapter aims to promote robotics in classroom by connecting robotics learning with various STEM curriculum standards.

Keywords Educational robotics · Constructionism · Maker movement in education · Technological literacy · Innovation literacy

A. Eguchi (✉)
Bloomfield College, Bloomfield, NJ, USA
e-mail: amy_eguchi@bloomfield.edu

© Springer International Publishing AG 2017
M.S. Khine (ed.), *Robotics in STEM Education*,
DOI 10.1007/978-3-319-57786-9_1

1.1   Introduction—Need for Change in Education

The speed of change in our society has been accelerating since the birth of the Internet. New technological tools are being introduced into our daily life more rapidly than ever before. Roughly twenty years ago, the cellular phone entered our lives just as the Internet started to connect personal computers. Now, the introduction of new iProducts and/or new smartphone products, such as Galaxy, occurs almost every six months. Creative project crowd-funding platforms, connected through the Internet, such as Kickstarter (http://www.kickstarter.com), Indiegogo (https://www.indiegogo.com/), and Quirky (https://www.quirky.com), are also contributing to the accelerated birth of innovative and creative technological tools by providing essential funding directly from potential and/or interested consumers.

Among these various technological advancements, the speed of the changes that robotics technology has created has been drastically increasing in recent years. News headlines from major news sources, including the New York Times, CNN, Wall Street Journal, and BBC, frequently featuring various robotic innovations, are a strong indication of such phenomenon. On March 10, 2004, the US Defense Advanced Research Projects Agency (DARPA) funded the first DARPA Grand Challenge held in a California desert, which was introduced as "a first-of-its-kind race to foster the development of self-driving ground vehicles" (OUTREACH@DARPA.MIL 2014). Its goal was to develop cars autonomously navigated a 142-mile course. The first DARPA Grand Challenge was followed by the second Grand Challenge in 2005, and the Urban Challenge in 2007, where autonomous cars navigated a complex course in a staged city environment. About a decade later, in 2014, Tesla, an American automaker, has already rolled out their first semiautonomous driving system, AutoPilot, in the market. In 2016, Tesla introduced all of their cars to be built with the necessary hardware for full self-driving capability.

On June 5, 2014, Softbank Mobile, a Japanese company, in collaboration with Aldebaran Robotics, a French company, unveiled *Pepper*, the world's first personal humanoid robot in Japan. Costing less than US$2000, Pepper is able to assist humans by reading and responding to human emotions (SoftBank Mobile Corp. and Aldebaran Robotics SAS 2014). Prior to Pepper, *NAO*, an autonomous and programmable humanoid robot developed by Aldebaran Robotics, has been used in various educational settings including RoboCup Soccer league for the development of algorithms for humanoid soccer since 2007, and for the research of children with Autism. Pepper has been developed to be "social companion for humans" (Middlehurst 2015, para. 5). The latest model is reported to have the ability to learn responses from a specific human using cloud technology-based AI (Tanabe 2015).

Governmental agencies from around the world have also invested in the development of robotics technologies. The DARPA funded the DARPA Robotics Challenge (DRC), which held between December 2013 to June 2015, ended with the DRC Finals. The DRC was "a competition of robot systems and software teams vying to develop robots capable of assisting humans in responding to natural and man-made disasters" (DARPA, n.a., para. 1). Its aim was to accelerate advanced research and development in robotics hardware and software that will enable future robots, in collaboration with humans, to perform the most hazardous activities in disaster zones, thus reducing casualties and saving lives. In February 2015, Japanese Ministry of Economy, Trade, and Industry has released its New Robot Strategy which outlines a plan to host the World Robot Summit in 2020, in the year when Tokyo Olympic is held (Ministry of Economy Trade and Industry, n.a.). This is part of the Japanese Economic Revitalization plan. In February 2016, the Executive Committee and the Advisory Board were formed to start the discussion and the planning for the World Robot Summit. The New Robot Strategy explains the goal of the World Robot Summit as a way to accelerate the research and development of robots and their introduction and diffusion into Japanese society. It aims to solve real issues in various areas, such as medical and health care, infrastructure inspection, agriculture, forestry and fisheries industry, manufacturing industry, service industry, and entertainment industry, use robotics technology as part of the competition, and demonstrate to society how robots can positively contribute to society. Furthermore, it aims to promote robots among people by introducing them through various games, and promoting various ways of living life with robots. The New Robot Strategy views the World Robot Summit as a driving force of the robot revolution, in which the infusion of robots will change the way people live every day.

The world and its economy are changing at such a rapid pace that it is impossible to predict what the economy will look like even at the end of next week (Robinson 2010). Despite all the drastic changes taking place in the world, public education has maintained almost the same system since its establishment in the middle of the nineteenth century (Robinson 2010). Although the requirements for an effective workforce have changed at the same speed as technological advancements, the majority of schools are continuing what was done in the past with little hope that they can adequately prepare students for the future (Robinson 2010). Some even state that if we could have teachers from the nineteenth-century time travel to our schools, they would have no problem teaching our students (Blikstien 2013). Because current public education places a heavy emphasis on memorization, our schools and curriculum are astonishingly similar to those in the nineteenth century (Blikstien 2013).

Long before the disconnect between the societal needs and what schooling provides students became an issue, Paulo Freire introduced his new view of education, leading to the development of the critical pedagogy approach. In his book, "Pedagogy of Oppressed," Freire points out that educational practice expects teachers to be narrators of facts and required students "to memorize mechanically the narrated content" (Freire 1994). In his view, students are turned into containers to be filled by the teacher. When using this *banking* approach to education (Freire 1994), students are required to receive, memorize, and repeat knowledge and/or facts that are provided by their teachers. Freire argues, "[t]he teacher talks about reality as if it were motionless, static, compartmentalized, and predictable. Or else he expounds on a topic completely alien to the existential experience of the students" (Freire 1994, p. 52). Freire warns:

> ...it is the people themselves who are filed away through the lack of creativity, transformation, and knowledge in this (at best) misguided system. For apart from inquiry apart from the praxis, individuals cannot be truly human. Knowledge emerges only through invention and re-invention, through the restless, impatient, continuing, hopeful inquiry, human beings pursue in the world, with the world, and with each other. (p. 53)

As society and economy changes, especially when changes in industries and workforce happen, "a new set of skills and intellectual activities become crucial for work, conviviality, and citizenship—often democratizing tasks and skills previously only accessible to experts" (Blikstien 2013, p. 1). Since, in previous era, the speed of the change was rather slower, the need for a new set of skills and intellectual activities did not happen as quick as it is now, maybe in every few centuries or so. This made it possible for public school system to successfully provide what society required. However, the recent acceleration of the change in society is forcefully requiring a new set of skills, intellectual activities, and ways of thinking to successful citizen while schooling has not made the transition to meet the needs.

In current society, more and more creativity and innovation are required. Yamakami (2012) emphasizes that recent technological advancement accelerated the speed of current innovation, drastically faster than in the past, and "as the speed of innovation has changed, the quantitative changes in speed have brought about qualitative changes in innovation" (p. 557). There is an urgent need for a radical and effective educational reform to keep up with societal changes. However, with the extensive focus on assessments through State mandate standardized testing, a concern is raised that more and more teachers are forced to teach to the test with more focus on memorization of facts. Even in the mid-1990s, similar concern as Freire was raised again (Grabinger and Dunlap 1995; Grabinger et al. 1997). Grabinger, Dunlap, and Duffield emphasized the importance of learning to think critically, to analyze and synthesize information in order to solve interdisciplinary problems, and to work collaboratively and productively with others in groups are important skills for participating effectively in society, and simply knowing some facts in a single domain and/or how to use tools are not enough for individuals to stay effective and competitive in increasingly complex society (Grabinger et al. 1997). Grabinger and Dunlap observed teachers in conventional classrooms and concluded that they used examples and problems that were simplified and decontextualized, leading to inadequate understanding of the acquired knowledge and an inability to effectively apply the knowledge to real situations. Students are often presented knowledge in conjunction with problems and examples that are not relevant to them and their needs and have no connection to their real life. The problems that students are asked to solve make them wonder, "Why do I need to know this?" (Grabinger and Dunlap 1995, p. 7). Because their learning is decontextualized and they cannot make connection between their learning and their life, student learning becomes an exercise of memorization of the facts transferred from their teachers. Moreover, they are not taught to acquire skills essential for effective thinking and reasoning (Grabinger and Dunlap 1995).

The US Government also highlights the need for fostering the contextualized knowledge and skills necessary to solve complex problems that we face every day among our youth (U.S. Department of Education 2015).

In a world that's becoming increasingly complex, where success is driven not only by *what* you know, but also by what you *can do* with what you know, it is more important than ever for our youth to be equipped with the knowledge and skills to solve tough problems, gather and evaluate evidence, and make sense of information. These are the types of skills that students learn by studying science, technology, engineering, and math—subjects collectively known as STEM (para. 2).

The government has raised the issue of the need for change in how we approach teaching as well, with a special focus recommended to be placed on STEM education. In 2011, President Obama stated:

> The first step in winning the future is encouraging American innovation. None of us can predict with certainty what the next big industry will be or where the new jobs will come from. Thirty years ago, we couldn't know that something called the Internet would lead to an economic revolution. What we can do – what America does better than anyone else – is spark the creativity and imagination of our people. (The White House 2011, para. 1)

The need for a STEM educated workforce is highlighted in the report titled "Strategy for American Innovation (SAI)—Securing Our Economic Growth and Prosperity" and "Federal Science, Technology, Engineering, and Mathematics (STEM) Education: 5 Year Strategic Plan." Moreover, it is recognized that the need for STEM knowledge and skills will continue to grow in the future (Tanenbaum 2016). The reports emphasize that it is essential that the country focuses on STEM education, with the aim to improve K-12 education, enhance US students' engagement in STEM disciplines, and graduate every student from high school ready for college and career, by inspiring and preparing more students, including girls and underrepresented groups, to excel in STEM field (National Economic Council, Council of Economic Advisers and Office of Science and Technology Policy 2011). The future economy and core employments will be driven primarily by innovation largely derived from advances in science and engineering (Committee on Highly Successful Schools for Programs for K-12 STEM Education Board on Science and Division of Behavioral and Social Sciences and Education 2011). However, the evidence shows that current education does not prepare a sufficiently large enough and well-equipped STEM workforce. For example, access to the full range of math and science courses that students need in order to pursue careers in STEM fields, such as Algebra I, geometry, Algebra II, calculus, biology, chemistry, and physics, is very limited in public schools. Although it is reported that more Asian-Americans and white students are likely to pursue STEM than other students of color, only 81% of Asian-American and 71% of white high school students attend high schools with a full range of math and science courses (U.S. Department of Education 2015). Furthermore, STEM literacy is necessary not only for those who pursue career in STEM fields but also for the general public.

Our current education system does not cultivate a culture of STEM, nor does it foster the development of a STEM literate public (U.S. Department of Education 2015). All students, no matter their race, zip code, or socioeconomic status, should be provided with the opportunity to be college-ready with STEM fluency. "STEM 2026" report also emphasizes the inequities in access, participation, and success in STEM subjects (U.S. Department of Education and Office of Innovation and Improvement 2016). They report the existence of persistent inequities between races, genders, socioeconomic groups, and among students with disabilities, thereby keeping the educational and poverty gaps wide and preventing us from fulfilling the needs of our technologically driven society. It is stated that effective STEM education accessible to and inclusive of all students is increasingly important so that our youth are equipped with "a new set of core knowledge and skills to solve difficult problems, gather and evaluate evidence, and make sense of information they receive from varied print and, increasingly, digital media" and prepared to become "a workforce where success results not just from what one knows, but what one is able to do with that knowledge" (U.S. Department of Education and Office of Innovation and Improvement 2016, p. i).

Similar issues exist in other countries. For example, the European Commission identified the STEM skill gap in participating countries (Communication from the Commission to the European Parliament, the Council and The European Economic and Social Committee and the Committee of the Regions 2011; Directorate-General for Research and Innovation 2015; Innovation Union 2015). It was pointed out that there are very few female students interested in science and pursue advanced level courses, even though innovation is required in both STEM and non-STEM related fields as well as in various aspects of our life (Communication from the Commission to the European Parliament, the Council, and the European Economic and Social Committee and the Committee of the Regions 2011). It was reported that one-third of the member countries had implemented awareness programs aiming to attract female students to STEM fields and research (Directorate-General for Research and Innovation 2015).

The skills to innovate cannot be cultivated through current educational practice focusing heavily on the memorization of knowledge without providing opportunities for students to transfer them into practice. There are urgent calls for innovative educational approaches worldwide that can foster skills for innovators including critical thinking, problem-solving, creativity, inventiveness, collaboration and teamwork, and communication skills through transdisciplinary, learner-centered, collaborative, and project-based learning. This chapter explains how educational robotics as a learning tool can create an effective learning environment to foster the learning of STEM and skills for innovators. In addition, it aims to make learning with robotics more accessible to students in every classroom by providing resources and making connection between learning with robotics and different learning standards.

1.2 Robotics in Education

Educational robotics or robotics in education is the phrase widely used to describe the use of robotics as a learning tool in the classroom. Popular interest in robotics has increased at astonishing rate not only in general society but even more so in the educational community in last several years (Benitti 2012). At the same time, robotics technology, once accessible only to experts and robotics scientists, has been more and more accessible for teachers and students of any age (Cruz-Martin et al. 2012; Mataric 2004), accelerated by the development of robotics components and tools suitable for school-age children (Eguchi and Almeida 2013). Large offerings of robotics materials prompted their rapid dissemination across all K-12 grades, mainly in robotics extracurricular activities (Benitti 2012; Eguchi 2007b; Sklar and Eguchi 2004). Even much younger students than in previous years now participate in educational robotics activities. Pre-K to kindergarten students now use robotics tools, such as KIBO by KinderLab Robotics (http://kinderlabrobotics.com), Dash and Dot by Wonder Workshop (https://www.makewonder.com), and BeeBot by Terrapin Software (https://www.bee-bot.us).

One of the most accessible robotics kits suitable for elementary to middle school students, LEGO Mindstorms (http://www.mindstorms.lego.com/), has been in the market for almost two decades. LEGO Mindstorms EV3, which came out in the summer of 2012, has more sophisticated controller and sensors than its precedent, NXT, and costs about $400. Many third-party sensors students can add on to the Mindstorms controllers, making the kit more robust and attractive to users. A German-based company, Fischertechnik (http://www.fischertechnik.de/en/), manufactures the ROBO set with similar features to the LEGO Mindstorms kit. Robotis (http://www.en.robotis.com), a Korean-based robotic company that produces DARwIn-OP, a humanoid robot, also makes educational robotics kits for school-age children. Their OLLO is for younger students (elementary school), and BIOLOID is for older students (upper elementary, middle school to high school). Daisen (http://daisen-netstore.com), a Japanese electronics company, has several robotic kits including the e-Gadget series that can be expanded with more robust sensors and motors.

The cost of the robotics kits has been one of the factors in the past that prevented the implementation of robotics in classroom. However, the price of robotics components has become more affordable in recent years. Arduino (https://www.arduino.cc) and Raspberry Pi (https://www.raspberrypi.org) are easy to use controller boards/microcomputers for middle to high school students. Arduino UNO Rev 3, one of Arduino series, costs $25. Raspberry Pi Foundation introduced Raspberry Pi ZERO at the end of 2015 for $5. Its latest version, Raspberry Pi 3, is $35. The rapid development of affordable controller boards and microcomputers is making handmade robotics/robotics maker activities more accessible for school-aged children (Fig. 1.1). These are just a few examples out of a very long list of robotics kits and components that are available for educators and students.

Fig. 1.1 Humanoid
performance robot created
with various controllers,
motors, and sensors
(© The RoboCup Federation.
Used with permission)

1.2.1 Foundation of Robotics in Education

Educational robotics has its roots in constructionism theory. The theory of constructionism became reality through Logo, a computer programming language for children, which in turn became the foundation for the development of the *Programmable Brick* for the LEGO Mindstorms (Martin et al. 2000). Constructionism theory was developed by Saymour Papert, a student of Jean Piaget. Papert built on to Piaget's constructivist theory to develop his constructionist theory. Constructivism theory highlights that:

> Knowledge is not a commodity to be transmitted. Nor is it information to be delivered from one end, encoded, stored and reapplied at the other end. Instead, knowledge is experience, in the sense that it is actively constructed and reconstructed through direct interaction with the environment. (Ackermann 1996, p. 27)

Children continuously construct new knowledge through their active interaction with the world while they try to make sense of it. A simple or direct instruction of facts or knowledge does not *stick* with children since knowledge is not constructed by them. It is suggested that children's knowledge construction needs to be supported by the manipulation of artifacts—*object to think with*. Piaget explains that learning involves constructing new knowledge out of prior knowledge by manipulating artifacts and observing their behavior (Piaget 1929, 1954).

To enhance children's learning, it is crucial to provide them opportunities where they can "engage in hands-on explorations that fuel the constructive process" (Ackermann 2001, pp. 1–2). In addition, since children construct knowledge through their interactions with the world—their own experience contributes to their understanding of the world—they develop a good reason to hold onto their views and understanding. Piaget points out that children rarely let go of their view of the world even when adults tell them that their view is wrong. Instead of providing the right knowledge, we should encourage them to continue to explore, express, and exchange their views (Ackermann 2004). What educational robotics can provide is the electronic manipulative or tool to explore, think, and interact with for children.

While the focus of constructivism is epistemology—theory of knowledge—constructionism focuses on *learning*. Papert explains:

Constructionism – the N word as opposed to the V word – shares constructivism's connotation of learning as "building knowledge structures" irrespective of the circumstances of the learning. It then adds the idea that this happens especially felicitously in a context where the leaner is consciously engaged in constructing a public entity, whether it's a sand castle on the beach or a theory of the universe. (Papert and Harel 1991)

The focus of constructivism is on the construction of knowledge in one's head. Whereas, constructionism focuses more on "the role of constructions in the world as a support for those in the head" (Bers 2008, p. 13), which is supported by physical or concrete construction *in the real world*. It is also highlighted that what differentiate constructionist learning from constructivism learning is its focus on "learning by constructing knowledge through the act of making something shareable" (Martinez and Stager 2013, p. 21). With constructionist learning, the *object to think with* is built or made, and what is physically constructed can be publicly shared—shown, discussed, examined, and admired by children. Children continuing this process of knowledge construction with physical artifacts will provide more learning experience that will be built upon, revising or reconstructing previously constructed knowledge (Papert 1993). Papert believes that the externalization of one's inner feelings and ideas contributes to the knowledge construction (Ackermann 2004). By expressing and sharing ideas in some tangible form, children shape and sharpen their ideas further. What educational robotics provides is a manipulative that a child can, not only think with, but also make her abstract idea and understanding of her world in her head *real*.

Educational robotics is a learning tool that allows or rather encourages children's exploration of their ideas using a technically and computationally enhanced tangible object. Papert emphasizes the importance of children's construction of knowledge through a tangible object by exploring their ideas. Eleanor Duckworth, another student of Piaget, echoes Papert's constructionist theory with her pedagogy of critical exploration. The critical exploration encourages children to actively and reflectively engage in a subject with their *wonderful ideas* which they developed through inquiry, and explore their ideas further through encountering complex materials and/or confusions, and trying out multiple possibilities, leading to the construction of new knowledge (Cavicchi et al. 2009). It denies direct teaching by adults or experts. Instead of providing answers or even implying that there is a *correct* answer, it emphasizes that teachers facilitate the personal process of learning. Duckworth states that the having of wonderful ideas is the essence of intellectual development (Duckworth 2006). Critical exploration considers learners as explores, like the ones in history, to face the areas of unknown, take risks, experience unexpected discoveries, through their journey of learning (Cavicchi et al. 2009). In addition, it considers a teacher as an explorer whose focus is on understanding of the individual child's experience of the exploration of his wonderful ideas and the process of the knowledge construction. Duckworth explains Piaget's basic idea of assimilation as a process:

> that a person takes any experience into her own previous understanding (schemes, structures); that we cannot assume that an experience whose meaning seems clear to us will have the same meaning to someone else. (Duckworth 2006, p. 158)

She emphasizes that a teacher must be determined to understand the meaning that any particular experience holds for her students (Duckworth 2005). Critical exploration suggests that children should be encouraged to think about, reflect on, and share their experience, while teachers should try to understand their thinking and understanding of the experience. Duckworth points out that critical exploration happens in two ways—"exploration of the subject matter by the child (the subject or the learner) and exploration of the child's thinking by adult (the researcher or the teacher)" (Duckworth 2005, p. 259).

Robots naturally spark children's interests and curiosity. As a learning tool, educational robotics excites children to explore their ideas through their inquiries and try out their hypotheses. It provides children with multiple ways to explore their wonderful ideas, find new discoveries, and build their knowledge through real-world experiences while using a technologically and computationally enhanced tool. The instantaneous feedback it provides, when children's ideas are tested on a robotic tool, can challenge and inspire them to explore their ideas further. In the following section, we will examine educational robotics and its role in the context of maker movement in education.

1.2.2 Educational Robotics in the Context of Maker Movement in Education

Making has been part of our culture since the beginning of human existence. Mark Hatch, the CEO of TechShop, explains:

> Making is fundamental to what it means to be human. We must make, create, and express ourselves to feel whole. There is something unique about making physical things. Things we make are like little pieces of us and seem to embody portions of our soul. (Hatch 2014, p. 11)

He argues that there is something about physical making that provides us with more personal fulfillment than virtual making. It is "its tangibility; you can touch it and sometimes smell and taste it" that give us satisfaction (p. 12).

The maker movement has gained increasing attention not only from the media and public, but also from the US Government. President Obama declared the first-ever White House Maker Faire to be hosted in June 2014, annually thereafter in June during the National Week of Making, and launched the Nation of Makers initiative (Kalil and Miller 2014). President Obama stated in 2015:

Makers and builders and doers—of all ages and backgrounds—have pushed our country forward, developing creative solutions to important challenges and proving that ordinary Americans are capable of achieving the extraordinary when they have access to the resources they need. During National Week of Making, we celebrate the tinkerers and dreamers whose talent and drive have brought new ideas to life, and we recommit to cultivating the next generation of problem solvers. (Kalil and Miller 2014, para. 3)

Original Maker Faire was launched in Bay Area in 2005. Now, there are two flagship events—Maker Faire Bay Area and World Maker Faire New York—in addition to the National Maker Faire and mini Maker Faires organized nationally and worldwide. The maker movement has been created organically by makers, builders, and doers from around the world. The maker movement is becoming a driving force, in combination with creative makers and innovative technologies, including lower priced and accessible microcontrollers and personal 3D printers, to accelerate the innovation in manufacturing, engineering, industrial design, medicine, hardware and software technologies, and education (Maker Faire, n.a.). Although *making* refers to any form of physical making or building, the *making* in the maker movement refers to the ones enhanced with digital and technological tools, including robotics.

In Maker Movement Manifesto, Hatch explains that there are nine ideas that characterize the maker movement: make, share, give, learn, tool up (make tools of making including the digital and technological tools accessible), play, participate, support, and change (Hatch 2014). Maker movement is also considered to have its foundation in constructionism (Martinez and Stager 2013) since it provides opportunities for learning by making. Through making activities, people problem-solve and construct new knowledge.

Incorporating the maker movement into education has the potential to transform current teaching practices. Bringing maker movement discussions to education will provide us with a venue for rethinking the definition of learner, a learner and a learning environment (Halverson and Sheridan 2014), and how to help our students *learn*. Making activities can potentially provide a series of activities that can help improve children's construction and reconstruction of knowledge, and empower them to actively engage in their learning. Moreover, it has a potential to change our STEM learning through the process of technologically and digitally enhanced making. Making can be organized with a set of activities designed with a various set of learning goals, enabling teachers to provide a transdisciplinary approach of learning. Since it is rooted in constructionism, maker activities focusing on the learning process engage the intersection of STEM learning, involving computer science, design, art, and engineering (Halverson and Sheridan 2014). Moreover, digital making can provide students with opportunities to make *real* powerful and wonderful ideas and use expressive tools (Blikstien 2013). Similar to robotics technologies, digital fabrication technology has become more and more accessible in recent years, which made it possible for the maker movement to enter classrooms, which in turn has made "the intellectual activities enabled by the new technology become more valued and important in classroom" (Blikstien 2013, p. 2).

Robotics in education is a pioneer of the maker movement in education since it has been in practice long before the maker movement emerged. Educational robotics has enhanced existing learning activities with a powerful technologically empowered medium (Blikstien 2013). The programmable brick or controller, connected with motors and various sensors, adds "computational behaviors to familiar materials—craft, LEGOs, wheels" (pp. 6–7). It provides children a new way of expressing their wonderful and powerful ideas with physical artifacts with technologically fueled everyday materials. Through the process and experience children have working with technologically enhanced artifacts, children construct and reconstruct their knowledge. The emergency of digital fabrication tools accessible to non-experts including children has powered up the capacity and quality of work that children can engage in. The digital fabrication has added another form of expression and realization of children's wonderful and powerful ideas to educational robotics (Figs. 1.2 and 1.3).

Advocates of the maker movement in education strongly suggest that it is important to make the maker learning opportunities accessible to all students, including females and underprivileged (Blikstien 2013; Halverson and Sheridan 2014; Martinez and Stager 2013) because it has a potential to attract students who are traditionally disengaged from technologically enhanced making activities. Lessons in educational robotics tend to start with building a robotics car. This is a great way of learning programming with motors and sensors. However, not everyone is attracted and/or interested in building a car (Rusk et al. 2008). This approach tends to attract tech-savvy boys. But girls may feel intimidated or shy

Fig. 1.2 Robotics creation (© The RoboCup Federation. Used with permission)

Fig. 1.3 Soccer robots
and performance robots
(© The RoboCup Federation.
Used with permission)

away from robotics activities just because it is a car-based robotics. Technologically enhanced maker activities that use everyday materials such as craft materials, fabrics, and various construction materials can lower the bar for those who think robotics is only for tech-savvy boys to jump in.

To make technologically enhanced maker education accessible to *all* learners, it is important to make sure that the materials and tools are accessible to *all* children. Recent development of cutting-edge technological tools, both software and hardware, has provided opportunities for children to engage in various technologically enhanced making activities such as "advanced scientific exploration, create interactive textiles, build simulations and games, program videogames, design virtual robotics system, create sophisticated 3D worlds and games though programming, build new types of cybernetics creatures, explore environmental science and geographical information systems," (Blikstien 2013, p. 5) and build robotics inventions. Although such developments have contributed to the popularity of maker movement and digital fabrication, there still is a divide in the population of potential users between the *haves* and *have-nots*. It is crucial to bring maker education into all classrooms so that everyone has a chance to learn from maker activities. Halverson and Sheridan point out that the maker movement reaches out to both formal and informal education and can potentially bring equity in education (Halverson and Sheridan 2014). We need to democratize the making in education for *all*. Halverson and Sheridan (Halverson and Sheridan 2014) emphasize:

> the great promise of the maker movement in education is to democratize access to the discourses of power that accompany becoming a producer of artifacts, especially when those artifacts use twenty-first-century technologies. (p. 500)

Technologically enhanced making in education can provide students with powerful tool for empowerment by enabling them with skills and knowledge important for their future as well as for making changes in society. To successfully bring maker education, especially educational robotics, into formal educational settings, teachers and educators interested need to be informed what students are learning through maker activities and what works. Because of this reason, it is important to identify the student learning outcomes through robotics making activities and how to make it work in formal classrooms. In the following section, various skills and learning outcomes that making with robotics can bring to classroom will be discussed.

1.3 Learning with Educational Robotics—Skills and Student Learning

Educational robotics is an effective tool for facilitating students' STEM learning. Studies show that learning with robots provides opportunities for students to obtain both content knowledge in physics, biology, geography, mathematics, science, electronics, and mechanical engineering, and acquiring critical academic skills, such as writing, reading, research, creativity, collaboration, critical thinking, decision-making, problem-solving, and communication skills, and design and computational thinking skills (Dimitris et al. 2009; Atmatzidou and Demetriadis 2012; Benitti 2012; Carbonaro et al. 2004; Eguchi 2014, 2016; Elkind 2008; Kolberg and Orlev 2001; Miller et al. 2008; Nourbakhsh et al. 2004; Oppliger 2002; Sklar and Eguchi 2004; Sklar et al. 2002, 2003).

Making with robotics is effective because it creates a fun and engaging hands-on learning environment for students. Unlike a traditional classroom setting where students listen to teacher's instruction in a more disciplined and structured way, making with robotics necessitates that students engage in manipulating, assembling, and reassembling materials while going through the design learning process and problem-solving program errors through trial and error. The learning approaches that enhance learning through making with robotics are project-based, problem-based, learning by design, student-centered, and constructionist learning approaches where the focus is on the process of learning, rather than the final product.

When working on making with robotics activities, it is highly recommended that students work in small groups (Eguchi 2012, 2015; Eguchi and Uribe 2012). By working in groups, students obtain skills needed for effective collaboration. The students are excited and motivated to share their ideas, engage in collaborative decision-making, provide constructive criticism, and acquire communication skills (Eguchi 2007a, c; Miller et al. 2008). Group work provides students the opportunity to explore and solve real-world problems with their peers. Working on a team-based and project-based robotics project helps students with low esteem to improve their technology capacity, teamwork skills, and communication skills (Miller et al. 2008).

Moreover, students working in groups on robotics making projects learn subject knowledge and the skills necessary for them to successfully complete a project while exploring real-world problems and challenges. As students tackle real-world problems, develop solutions, and demonstrate their learning by physically testing their solutions with their robotics creation, they are actively engaged in a kind of learning that results in deeper subject knowledge acquisition (Edutopia, n.a.). Making with robotics makes possible a transdisciplinary learning environment where students can come across various concepts in STEM and other disciplines in a contextualizing fashion. With contextualized learning while making robotics, abstract concepts, such as friction and momentum, become visible and concrete for students to grasp as they try out their ideas with their robotics invention (Blikstien 2013) (Fig. 1.4).

There are new literacies, including mathematics, engineering, science (Blikstien 2013) as well as technological and innovation literacies, considered to be crucial for students to gain fluency in the twenty-first-century skills necessary for becoming effective citizens of the future. Learning these new literacies is supported by making with robotics. Technological and innovation literacies will be further introduced below, followed by various student learning outcomes supported by making with educational robotics activities.

1.3.1 Technological Literacy

As various technological tools have become more and more advanced and accessible to non-experts including children, the intellectual activities and learning through these activities have become more valued by society and becoming increasingly acknowledged by the education community (Blikstien 2013). At the same time, technological skills, such as typing, have become less valued and technological fluency has become more valued. For example, desired computer skills have become *computational fluency* or *literacy* (diSessa cited in Blikstien 2013), and a broader understanding of technological fluency has been expanded to include engineering knowledge and the process of engineering design (National Research Council cited in Blikstien 2013). Computational thinking is also an important part of technological literacy. Blikstien explains that technological

Fig. 1.4 Students collaborating on programming a soccer robot (© The RoboCup Federation. Used with permission)

literacy is a "general set of skills and intellectual dispositions for all citizens," while technological competence means "in-depth knowledge that professional engineers and scientists need to know to perform their work" (p. 3). Being fluent in technology is now desirable for everyone in society. Technologically enhanced making with robotics contributes to the development of technological fluency among students.

1.3.2   Innovation Literacy

Innovation literacy, another set of crucial skills that our children need to acquire, has been introduced and supported in several publications from various fields (i.e., Erdogan et al. 2013; Gelb and Caldicott 2007; Yamakami 2012). Erdogan et al. (2013) urge educators to consider innovation to be a necessary focus of student learning of twenty-first-century skills (Partnership for 21st Century Skills 2008), requiring the development of new learning environments that foster innovation. It is suggested that students learn innovation through contextualized transdisciplinary approaches since innovation requires a range of skills and knowledge for bringing innovative ideas into reality (Govindarajan cited in Gelb and Caldicott 2007).

Erdogan et al. (2013) describe innovation literacy as an interdisciplinary literacy involving reading, math, and science literacies, and social skills such as collaboration and originality. Innovation literacy is a set of skills that enables people to understand and use information, such as texts and graphs, and to make logically and scientifically supported decisions on how to develop innovative outcomes and/or solutions (Erdogan et al. 2013). OECD (2015) supports this view of innovation literacy. Innovation literacy includes both domain-specific skills and knowledge and broader competencies such as creativity, critical thinking, collaboration/teamwork, and communication skills. The Conference Board of Canada addresses the skills for innovation including creativity, problem-solving, continuous improvement (persistence) skills, risk assessment and risk-taking skills, and relationship—building and communication skills (The Conference Board of Canada, n.a.).

Making with robotics fosters various innovation literacy skills including subject-related literacy and academic skills such as engineering design including continuous improvement skills, computational thinking, creativity, problem-solving, communication and collaboration skills, and creativity, as explained in the previous sections.

1.3.3   Student Learning with Educational Robotics

Educational robotics is a learning tool that fosters various skills and knowledge essential for *every* student to take part in creating the future innovations that society needs. This skill set and knowledge will also enable students to turn their imagination and innovation into reality as well as find a new way of self-expression (Alimisis 2013). However, for school-age children, most making with robotics activities has been part of informal education (Alimisis 2013; Benitti 2012; Blikstien 2013; Eguchi 2007b; Sklar and Eguchi 2004). There are several factors preventing teachers from bringing robotics into their classroom. Robotics or engineering is not a core curriculum strand in most schools. There have been enthusiastic and creative teachers who have developed ideas for ways to connect learning through robotics making with traditional subject matter learning outcomes and standards. It is difficult to fit making with robotics into existing school curriculum which defines learning by subject areas, disconnected from each other. Although it is clear that contextualized transdisciplinary approaches create optimal learning opportunities for technological and innovative literacies as well as meaningful knowledge acquisition, current curriculum and classroom practice do not embrace the potential making with robotics can bring into classroom. When trying to bring making with robotics into formal education, it becomes crucial to address the student learning outcomes that align with curriculum standards.

There are several pioneer teachers who aligned and addressed making with robotics activities with student learning outcomes (i.e., Bratzel 2007, 2009, 2014; Kee 2011, 2013, 2015, 2016). Their efforts have helped some of the early adaptors to bring making with robotics into classroom. More efforts to connect making with robotics activities with various learning standards set by governmental agencies have been made by various robotics competitions. They have responded to educators' needs to align activities with student learning outcomes so that more teachers will be able to adopt robotics making in their curriculum. For example, VEX and VEX robotics Autodesk's one-semester robotics curriculum for 9–12 grades, which is aligned with four curriculum standards—ITEEA standards for Technological Literacy, Common Core State Standards for Mathematics and English, Next Generation Science Standards (VEX EDR, n.a.). FIRST Robotics Competition also provides its FIRST Robotics Competition Standard Alignment Map aligning their activities with four curriculum standards—Common Core State Standards, Next Generation Science Standards, Next Generation Science Standards, and 21st Century Learning Skills (FIRST Robotics Competition 2016). RoboCupJunior Australia, an Australian division of RoboCup Junior, prepared curriculum map to address Australian Curriculum—Technologies set by ACARA (Australian Curriculum, Assessment and Reporting Authority) (Moreton et al. 2014).

In the following section, we will focus on some of the US Government mandated standards and describe how making with robotics activities can support students as they work to achieve their learning goals.

1.3.3.1 Mathematics with Educational Robotics

Since educational robotics is a tool used to promote STEM learning through hands-on activities, mathematics is one of the subject areas addressed through the making with robotics activities. Common Core State Standards for Mathematics presents eight Mathematics Practical Standards that teachers of all levels should aim to develop in their students:

- MP1: Make sense of problems and persevere in solving them
- MP2: Reason abstractly and quantitatively
- MP3: Construct viable arguments and critique the reasoning of others
- MP4: Model with mathematics
- MP5: Use appropriate tools strategically
- MP6: Attend to precision
- MP7: Look for and make sense of structure
- MP8: Look for and express regularity in repeated reasoning

Common Core State Standards Initiative (2010).

Making with robotics activities addresses all eight Mathematics Practice Standards. While engaged in making with robotics activities, students develop skills to dissect, understand, and analyze problems that they encounter, then develop, test, and improve solutions using data collected and mathematical formulas (MP1). By solving a variety of problems that they encounter while working on robotics making —designing, building, and programming their robotics creations, students develop the skill to think, understand, and solve problems abstractly and quantitatively (MP2). Since making with robotics uses the project-based approach whereby a small group of students work together on their robotics creation, students will develop communication and collaboration skills including constructing viable arguments and critiquing the reasoning of others in constructive ways (MP3). In the process of making, students create different solutions for their construction models and codes using various mathematical tools (i.e., graphs, charts, and tables) for decision-making to improve their robotics creation (MP4). Students engaged in making with robotics activities learn to select the appropriate tool needed to solve problems that they encounter. For example, students examine the problem that they have to solve, evaluate possible solutions, and select the correct sensor for the best solution that they choose (MP5). Since students work on their robotics creation as a group, good communication is a key to success. Students develop skills to communicate precisely to each other. Through their robotics work, students learn to use academic language to precisely communicate their ideas with details (MP6). While building and programming their robotics creations, students learn to recognize and use structures and patterns (MP7). Students engaged on robotics making activities go through a reiterated process as they solve the problems and challenges that they face. Throughout the process, they learn to look for and express regularity in repeated reasoning (MP8). Through the robotics making activities, students encounter a variety of occasions when they apply mathematical concepts, such as the concepts of number and operations, measurement and data, geometry, ratios and proportional relationships, expressions, and equations. Making with robotics not only provides students opportunities to learn mathematical concepts but also applies their learning to a real-world situation.

1.3.3.2 English Language Arts with Educational Robotics

Common Core State Standards for English Language Arts (ELA) emphasize the importance of the College and Career Readiness Anchor (CCRA) Standards as they form the backbone of the ELA/literacy standards. Student learning through making with robotics activities addresses some of the CCRA standards, which are listed below:

- CCRA/R1: Read closely to determine what the text says explicitly and to make logical inferences from it; cite specific textual evidence when writing or speaking to support conclusions drawn from the text
- CCRA/R4: Interpret words and phrases as they are used in a text, including determining technical, connotative, and figurative meanings, and analyze how specific word choices shape meaning or tone
- CCRA/R7: Integrate and evaluate content presented in diverse formats and media, including visually and quantitatively, as well as in words
- CCRA/W1: Write arguments to support claims in an analysis of substantive topics or texts, using valid reasoning and relevant and sufficient evidence
- CCRA/W2: Write informative/explanatory texts to examine and convey complex ideas and information clearly and accurately through the effective selection, organization, and analysis of content
- CCRA/W4: Produce clear and coherent writing in which the development, organization, and style are appropriate to task, purpose, and audience
- CCRA/W5: Develop and strengthen writing as needed by planning, revising, editing, rewriting, or trying a new approach
- CCRA/W6: Using technology, including the Internet, to produce and publish writing and to interact and collaborate with others
- CCRA/W7: Conduct short as well as more sustained research projects based on focused questions, demonstrating understanding of the subject under investigation
- CCRA/W8: Gather relevant information from multiple print and digital sources, assess the credibility and accuracy of each source, and integrate the information while avoiding plagiarism
- CCRA/W9: Draw evidence from literary or informational texts to support analysis, reflection, and research
- CCRA/W10: Write routinely over extended time frames (time for research, reflection, and revision) and shorter time frames (a single sitting or a day or two) for a range of tasks, purposes, and audiences

- CCRA/SL1: Prepare for and participate effectively in a range of conversations and collaborations with diverse partners, building on others' ideas and expressing their own clearly and persuasively
- CCRA/SL2: Integrate and evaluate information presented in diverse media and formats, including visually, quantitatively, and orally
- CCRA/SL3: Evaluate a speaker's point of view, reasoning, and use of evidence and rhetoric
- CCRA/SL4: Present information, findings, and supporting evidence such that listeners can follow the line of reasoning, and the organization, development, and style are appropriate to task, purpose, and audience
- CCRA/SL5: Make strategic use of digital media and visual displays of data to express information and enhance understanding of presentations
- CCRA/SL6: Adapt speech to a variety of contexts and communicative tasks, demonstrating command of formal English when indicated or appropriate
- CCRA/L1: Demonstrate command of the conventions of Standard English grammar and usage when writing or speaking
- CCRA/L2: Demonstrate command of the conventions of Standard English capitalization, punctuation, and spelling when writing
- CCRA/L4: Determine or clarify the meaning of unknown and multiple-meaning words and phrases by using context clues, analyzing meaningful word parts, and consulting general and specialized reference materials, as appropriate
- CCRA/L5: Demonstrate understanding of figurative language, word relationships, and nuances in word meanings
- CCRA/L6: Acquire and use accurately a range of general academic and domain-specific words and phrases sufficient for reading, writing, speaking, and listening at the college and career readiness level; demonstrate independence in gathering vocabulary knowledge when encountering an unknown term important to comprehension or expression

Making with robotics activities addresses the Common Core Standards for ELA —Science and Technical Subjects for grade level. Although educational robotics is a tool to promote STEM learning, the project-based approach requires students to engage in various reading and writing tasks. When students work on a project of their choice, students conduct research on the topic of their project, read and analyze materials, come up with a solution, and write it up. They use various media including digital media from the Internet for their research. Since it is a collaborative group work, students learn to communicate effectively to each other. They learn to evaluate, analyze, and understand each other's point of view, reasoning, and use of evidence. Throughout the process of making, students are required to keep reflective journals or log books, in which they record their learning journey including their problem-solving strategies, data that they collected and their analysis of the data, summary of their group discussion, problems that they faced and how they plan to solve it, mistakes they made and how to avoid it next time, and their feelings. In addition, students work on presenting their robotics creation to the public—classmates, teachers, families, and school and beyond. To prepare for the presentations, students work on evaluating and summarizing their making process using various writings that they have done through the process, synthesizing their own writing, various sources, and data used during the making process, into the final form of presentation. Students use any tools to create their presentation including digital media tools, visual display, and/or craft materials.

1.3.3.3 Engineering Design (Engineering Literacy) with Educational Robotics

The Next Generation Science Standards (NGSS) emphasizes engineering design practices as a skill set that is an integral part of science education, which all citizens should learn (Next Generation Science Standard 2013). NGSS places engineering design as the same level as scientific inquiry when teaching science disciplines in K-12 education. It considers engineering and science as instrumental in creating solutions to the major challenges we face. The key to the engineering design is the iterative cycle of the design process when used to solve problems. There are three core components in engineering design set by NGSS:

(a) *Defining and delimiting engineering problems* involves stating the problem to be solved as clearly as possible in terms of criteria for success, and constrains or limits.
(b) *Designing solutions to engineering problems* begins with generating a number of different possible solutions, then evaluating potential solutions to see which ones best meet the criteria and constraints of the problem.
(c) *Optimizing the design solution* involves a process in which solutions are systematically tested and refined and the final design is improved by trading off less important features for those that are more important (p. 2).

The three components do not always follow in order. Since the engineering design process is an iterative cycle, a problem can be redefined or new solutions can be generated to replace an idea that does not work at any stage of the cycle.

Engineering design process is one of the important skills that students obtain through the learning by design with robotics making. Students use trial and error strategy during the engineering design process to refine and recreate their solutions to the problems that they face. Students could redesign, reconstruct, and reprogram their robotics creations as many times as needed to come up with a satisfied final design and solution.

1.3.3.4 Computational Thinking with Educational Robotics

Computational thinking has gained great attention in the field of education in recent years, especially after the *Hour of Code* was launched in December 2013 in the USA and England implemented its computing education in 2014. In a seminal article on computational thinking by Jeannette Wing in 2006, Wing predicted that computational thinking would be a fundamental skill used by everyone in the world by the middle of twenty-first century (Wing 2006). Computational thinking has its focuses on the process of abstraction. Wing's view of computational thinking is as follows:

> [i]nformally, computational thinking describes the mental activity in formulating a problem to admit a computational solution. The solution can be carried out by a human or machine, or more generally, by combinations of humans and machines. (Wing 2010, para. 1)

Wing's term of computational thinking is broadly described as the design and analysis of problems and their solutions.

The International Society for Technology in Education (ISTE) and the Computer Science Teachers Association (CSTA) share the common understanding that computational thinking should be integral part of education for school-age children. SITE and CSTA highlight that "computational thinkers are the creators, designers, and developers of the technology tools and systems that are now contributing to major advances in almost every field of human understanding and endeavor" (Computer Science Teachers Association and International Society for Technology in Education 2011, p. 7). There is a greater pressure in society to educate more computational thinkers than ever before. Although computational thinking is not one of the standards required by government except for a couple of countries, like England, it is important to consider including it as part of student learning through making with robotics. ISTE and CSTA collaborated to develop the following operational definition of computational thinking for primary and secondary education:

Computational thinking (CT) is a problem-solving process that includes (but is not limited to) the following characteristics:

- Formulating problems in a way that enables us to use a computer and other tools to help solve them
- Logically organizing and analyzing data
- Representing data through abstractions such as models and simulations
- Automating solutions through algorithmic thinking (a series of ordered steps)
- Identifying, analyzing, and implementing possible solutions with the goal of achieving the most efficient and effective combination of steps and resources
- Generalizing and transferring this problem-solving process to a wide variety of problems

These skills are supported and enhanced by a number of dispositions or attitudes that are essential dimensions of CT. These dispositions or attitudes include the following:

- Confidence in dealing with complexity
- Persistence in working with difficult problems
- Tolerance for ambiguity
- The ability to deal with open-ended problems

- The ability to communicate and work with others to achieve a common goal or solution (Computer Science Teachers Association and International Society for Technology in Education 2011, p. 13)

Since computing is part of making with robotics, it provides the right environment in which students obtain computational thinking skills. For example, students demonstrate their abstraction and algorithmic thinking through the algorithm they create since an algorithm is an abstraction of a process, broken down in ordered steps. Such steps are created with sensor inputs, carry out the series of ordered steps, and produce outputs to accomplish the targeted goal. Students who can create effective algorithms for their problems develop the skill to formulate the steps in a way to effectively use the robotic tool. This requires the skills to identify, analyze, and implement the solution with the most effective and efficient steps. Experienced programmers can create effective but simple solutions. Those abilities need to be supported by the right dispositions, including persistence, tolerance, an ability to communicate and work effectively with others, and an ability to deal with open-ended problems. Such dispositions can be obtained from their participation in making with robotics activities and the learning process. Through maker activities with robotics, students gain the confidence needed to deal with complexity. Quite often, students encounter complex problems while making with robotics, which help students to develop the confidence to persist.

1.4 Conclusion

The speed change introduced into society, especially with technological domains, has been accelerating since the birth of the Internet. This means that a different set of skills, such as creativity and innovation, are required for the workforce to effectively continue to invent and innovate. The problem that we are facing is that our public education has not kept pace with the work force needs of our rapidly changing society. A new wave of innovation in education, such as educational robotics, maker movement, and digital fabrication, has the potential to bring about the necessary changes in formal education. Although educational robotics, maker movement, and digital fabrication are not new in its use by informal education in recent years, it is important to bring these approaches and tools into formal educational classroom settings so that such learning experiences are accessible to *all* students, not only those who are privileged or boys.

One way of lowering the barrier for teachers and educators is connecting such learning activities with existing learning standards. However, simply bringing making with robotics activities into classrooms does not automatically bring desirable learning outcomes. Since making with robotics has its base in constructionist learning, teachers have to create a constructionist learning environment where the focus is on learners' exploration of their ideas using technological tools. In other words, students become the agent to *program* the computer and robots rather than just a consumer of technology (Blikstien 2013). By doing so, students acquire "a sense of mastery over a piece of the most modern and powerful technology" (Papert 1993, p. 5). The technology provides a powerful tool for students to build "their own intellectual structures" (Papert 1993, p. 7). With learner-centered approaches, teachers have to refrain from traditional ways of *teaching* and become facilitators of students' learning. It is also necessary for students to change from *passive* learners to *active* learners. Since the core activities are *making*, the students' learning should naturally shift from learning by *listening* to learning by *doing*.

Teacher as a facilitator of students' learning requires various skills in teachers. One of the skills is the ability to scaffold students' learning by asking the right questions to bring students' own inquiries out. With making with robotics, there is generally no one right way to solve a challenge. Not having one *right* answer but multiple ways of tackling a problem is an experience with which many teachers are not familiar. Not having one correct answer tends to make both teachers and students uncomfortable. However, many times, students learn the most from discussions and teacher's penetrating questions (Rogers and Portsmore 2004). It is crucial for teachers to provide learning opportunities that are open for students' ideas and let students create and design technological making projects. However, young students may get stuck or lost in their own ideas (Bers 2008). It is important that teachers scaffold the process of making with guiding and provoking questions. Hmelo-Silver, Duncan, and Chinn suggest that scaffolding can provide students opportunities to "engage in complex tasks that would otherwise be beyond their current abilities" (Han and Bhattacharya 2001; Hmelo-Silver et al. 2007, p. 100). Scaffolding, includes providing coaching, modeling, guiding, task structuring, pushes students to think deeply, and supports them to become effective information seekers and problem-solvers, as well as expert at finding help and necessary resources for themselves (Bers 2008; Hmelo-Silver et al. 2007).

Making with robotics activities needs to have broader perspectives in order to be inclusive and responds to various student interests. If teachers use conventional educational robotics approaches such as using robotics car, they can reach only the children talented in science, math, and technology, and fewer girls than boys. Studies have shown that the way educational robotics is introduced into the educational settings is often unnecessarily narrow in its focus (Rusk et al. 2008). Rusk, Resnick, Berg, and Pezalla-Granlund suggest robotics making activities be designed to use a theme-based approach rather than challenge-focused approach, a transdisciplinary approach in order to connect various subject areas, especially art and engineering—STREAM (STEM with Robotics and Arts), or a storytelling/ narrative approach as a new way of self-expression. An end of unit exhibition provides an opportunity for students to share their robotic creations (Rusk et al. 2008). By widening unit outcomes, teachers can engage and encourage a wider diversity of student participation. Creating inclusive learning environments using making with robotics activities will attract students who may not self-identify as strong in mathematics and/or science, as well as girls who think robotics is only for boys.

By bringing learning through making with robotics into every classroom, educators have the potential to provide *all* students with the opportunity to learn the skills and knowledge that they need to become effective members of the workforce and future innovators and creators. Making with robotics provides transdisciplinary learning environments in which students can encounter a range of STEM concepts as well as concepts from other subject areas including English, arts, and history in contextualized fashion. It also fosters the learning of various new literacies including technological and innovation literacies. Making with robotics can provide non-traditional learning environment that sparks students' interests and imagination. It can inspire *all* students' curiosity, enthusiasm for learning, and build self-confidence (Rogers and Portsmore 2004). Making with robotics has the possibility to become a game-changer in education, turning traditional education into a new form of *innovative* learning experience for all students.

References

Ackermann, E. K. (1996). Perspective-taking and object construction: Two keys to learning. In Y. Kafai & M. Resnick (Eds.), *Constructionism in practice: Designing, thinking, and learning in a digital world* (pp. 25–37). Mahwah, New Jersey: Lawrence Erlbaum Associates.

Ackermann, E. K. (2001). *Piaget's constructivism, paper's constructionism: What's the difference?* pp. 1–11. Retrieved from http://learning.media.mit.edu/content/publications/EA. Piaget_Papert.pdf

Ackermann, E. K. (2004). Constructing knowledge and transforming the world. In M. Tokoro & L. Steels (Eds.), *A learning zone of one's own: Sharing representations and flow in collaborative learning environments* (pp. 15–37). Washington, DC: IOS Press.

Alimisis, D. (2013). Educational robotics: Open questions and new challenges. *Themes in Science & Teaching Education, 6*(1), 63–71.

Alimisis, D., & Kynigos, C. (2009). Constructionism and robotics in education. In D. Alimisis (Ed.), *Teacher education on robotics-enhanced constructivist pedagogical methods*. Athens, Greece: School of Pedagogical and Technological Education.

Atmatzidou, S., & Demetriadis, S. (2012). *Evaluating the role of collaboration scripts as group guiding tools in activities of educational robotics*. Paper presented at the 2012 12th IEEE International Conference on Advanced Learning Technologies, Rome, Italy.

Benitti, F. B. V. (2012). Exploring the educational potential of robotics in schools: A systematic review. *Computers & Education, 58*, 978–988.

Bers, M. U. (2008). *Blocks to robots: Learning with technology in the early childhood classroom*. New York, NY: Teachers College Press.

Blikstien, P. (2013). Digital fabrication and 'making in education": The democratization of invention. In J. W. H. C. Buching (Ed.), *FabLabs: Of makers and inventors*. Bielefeld, Germany: Transcript Publishers.

Bratzel, B. (2007). *Physics by design: RoboLab activities for the NXT and RCX*. Knoxville, TN: College House Enterprises LLC.

Bratzel, B. (2009). *Physics by design with NXT Mindstorms*. Knoxville, TN: College House Enterprises LLC.

Bratzel, B. (2014). *STEM by design: Teaching with LEGO Mindstorms EV3*. Knoxville, TN: College House Enterprises LLC.

Carbonaro, M., Rex, M., & Chambers, J. (2004). Using LEGO robotics in a project-based learning environment. *Interactive Multimedia Electronic Journal of Computer Enhanced Learning,* 6(1).

Cavicchi, E., Chiu, S.-M., & McDonnell, F. (2009). Introductory paper on critical explorations in teaching art, science, and teacher education. *The New Educator, 5,* 189–204.

Committee on Highly Successful Schools for Programs for K-12 STEM Education Board on Science, E. a. B. o. T. a. A., & Division of Behavioral and Social Sciences and Education. (2011). *Successful K-12 STEM education—Identifying effective approaches in science, technology, engineering, and mathematics.* Washington D.C.: The National Academies Press.

Common Core State Standards Initiative. (2010). *Standard for mathematical practice.* Retrieved from http://www.corestandards.org/Math/Practice/

Communication from the Commission to the European Parliament, the Council, & The European Economic and Social Committee and the Committee of the Regions. (2011). *Europe 2020 flagship initiative innovation union SEC (2010) 1161.* Luxembourg: Publications Office of the European Union.

Computer Science Teachers Association, & International Society for Technology in Education. (2011). *Computational thinking leadership toolkit.* Retrieved from http://www.iste.org/docs/ct-documents/ct-leadershipt-toolkit.pdf?sfvrsn=4

Cruz-Martin, A., Fernandez-Madrigal, J. A., Galindo, C., Gonzalez-Jimenez, J., & Stockmans-Daou, C. (2012). A LEGO Mindstorms NXT approach for teaching at data acquisition, control systems engineering and real-time systems undergraduate courses. *Computers & Education, 59,* 974–988.

DARPA. (n.a.). *DARPA robotics challenge finals 2015: Overview—What is the DARPA robotics challenge (DRC)?* Retrieved from http://www.theroboticschallenge.org/overview

Directorate-General for Research and Innovation. (2015). *State of the Innovation Union 2015.* Luxemburg: Publication Office of the European Union.

Duckworth, E. (2005). Critical exploration in the classroom. *The New Educator, 1*(4), 257–272.

Duckworth, E. (2006). *The having of wonderful ideas: and other essays on teaching and learning* (3rd ed.). New York, NY: Teachers College Press.

Edutopia. (n.a., 2008, February 28). *Project-based learning.* Retrieved from http://www.edutopia.org/project-based-learning

Eguchi, A. (2007a, March). *Educational robotics for elementary school classroom.* Paper presented at the Society for Information Technology and Education (SITE), San Antonio, TX.

Eguchi, A. (2007b). Educational robotics for elementary school classroom. In *Proceedings of the Society for Information Technology and Education (SITE),* pp. 2542–2549.

Eguchi, A. (2007c). Educational robotics for undergraduate freshmen. In *Proceedings of the World Conference on Educational Multimedia, Hypermedia and Telecommunications,* pp. 1792–1797.

Eguchi, A. (2012). Student learning experience through CoSpace educational robotics. In *Proceedings of the Society for Information Technology & Teacher Education International Conference.*

Eguchi, A. (2014). Why robotics in education? Robotics as a learning tool for educational revolution. In *Proceedings of the Society for Information Technology & Teacher Education International Conference.*

Eguchi, A. (2015). Educational robotics as a learning tool for promoting rich environments for active learning (REALs). In J. Keengwe (Ed.), *Handbook of research on educational technology integration and active learning* (pp. 19–47). Hershey, PA: Information Science Reference (IGI Global).

Eguchi, A. (2016). Computational thinking with educational robotics. In *Proceedings of the Society for Information Technology & Teacher Education International Conference.*

Eguchi, A., & Almeida, L. (2013). RoboCupJunior: Promoting STEM education with robotics competition. In *Proceedings of the Robotics in Education.*

Eguchi, A., & Uribe, L. (2012). Educational robotics meets inquiry-based learning. In L. Lennex & K. F. Nettleton (Eds.), *Cases on inquiry through technology in math and science: Systemic approaches*. Hershey, PA: Information Science Reference (IGI Global).

Elkind, D. (2008). Forward. In M. U. Bers (Ed.), *Block to robots* (pp. xi–xiv). New York, NY: Teachers College Press.

Erdogan, N., Corlu, M. S., & Capraro, R. (2013). Defining innovation literacy: Do robotics programs help students develop innovation literacy skills? *International Online Journal of Educational Sciences, 5*(1), 1–9.

FIRST Robotics Competition. (2016). *Standard alignment map*. Retrieved from http://www.firstinspires.org/resource-library/frc/standard-alignment-map

Freire, P. (1994). *Pedagogy of the oppressed* (30th ed.). New York, NY: Bloomsbury Academic.

Gelb, M., & Caldicott, S. M. (2007). *Innovate like Edison: The success system of America's greatest inventor*. New York, NY: Penguin Group.

Grabinger, S., & Dunlap, J. C. (1995). Rich environments for active learning: A definition. *Research in Learning Technology, 3*(2), 5–34.

Grabinger, S., Dunlap, J. C., & Duffield, J. A. (1997). Rich environment for active learning, in action: Problem-based learning. *Research in Learning Technology, 5*(2), 5–17. doi:10.1080/0968776970050202.

Halverson, E. R., & Sheridan, K. M. (2014). The maker movement in education. *Harvard Educational Review, 84*(4), 495–504.

Han, S., & Bhattacharya, K. (2001). Constructionism, learning by design, and project based learning. In M. Orey (Ed.), *Emerging perspectives on learning, teaching and technology*.

Hatch, M. (2014). *The maker movement manifesto—Rules for innovation in the new world of crafters, hackers, and tinkerers*. New York, NY: McGraw-Hill Education.

Hmelo-Silver, C. E., Duncan, R. G., & Chinn, C. A. (2007). Scaffolding and achievement in problem-based and inquiry learning: A response to Kirschner, Sweller, and Clark (2006). *Educational Psychologist, 42*(2), 99–107.

Innovation Union. (2015). *Promoting excellence in education and skills development*. Retrieved from http://ec.europa.eu/research/innovation-union/index_en.cfm?pg=action-points

Kalil, T., & Miller, J. (2014, February 3). *Announcing the first white house maker faire*. Retrieved from http://www.whitehouse.gov/blog/2014/02/03/announcing-first-white-house-maker-faire

Kee, D. (2011). *Classroom activities for the busy teacher: NXT* (2nd ed.). CreateSpace Independent Publishing Platform.

Kee, D. (2013). *Classroom activities for the busy teacher: EV3*. CreateSpace Independent Publishing Platform.

Kee, D. (2015). *Classroom activities for the busy teacher: VEX IQ with Modkit for VEX*. CreateSpace Independent Publishing Platform.

Kee, D. (2016). *Classroom activities for the busy teacher: VEX IQ with ROBOTC Graphical*. CreateSpace Independent Publishing Platform.

Kolberg, E., & Orlev, N. (2001). *Robotics learning as a tool for integrating science-technology curriculum in K-12 schools*. Paper presented at the 31st ASEE/IEEE Frontiers in Education Conference, Reno, NV.

Maker Faire. (n.a.). *The maker movement*. Retrieved from http://makerfaire.com/maker-movement/

Martin, F., Mikhak, B., Resnick, M., Silverman, B., & Berg, R. (2000). To Mindstorms and beyond: Evolution of a construction kit for magical machines. In A. Druin & J. Hendler (Eds.), *Robots for kids: Exploring new technologies for learning* (pp. 9–33). San Diego, CA: Academic Press.

Martinez, S. L., & Stager, G. (2013). *Invent to learn: Making, tinkering, and engineering in the classroom*. Torrance, CA: Constructing Modern Knowledge Press.

Mataric, M. J. (2004). *Robotics education for all ages*. Paper presented at the American Association for Artificial Intelligence Spring Symposium on Accessible, Hands-on AI and Robotics Education. http://robotics.usc.edu/~maja/publications/aaaissymp04-edu.pdf

Middlehurst, C. (2015, November 2). 'Human' robot Pepper proves popular again and sells out in less than a minute in Japan. *The Telegraph*. Retrieved from http://www.telegraph.co.uk/news/worldnews/asia/japan/11969300/Human-robot-Pepper-proves-popular-again-and-sells-out-in-less-than-a-minute-in-Japan.html

Miller, D. P., Nourbakhsh, I. R., & Sigwart, R. (2008). Robots for education. In B. Siciliano & O. Khatib (Eds.), *Springer handbook of robotics* (pp. 1283–1301). New York, NY: Springer New York, LLC.

Ministry of Economy Trade and Industry. (n.a.). *The executive committee and the advisory board for the international robot competition meeting (1st) related documents*. Retrieved from http://www.meti.go.jp/committee/kenkyukai/seizou/robot_competition/001_haifu.html

Moreton, B., Elias, G., Bowler, S., Tardiani, G., & Kee, D. (2014). *ACARA Link*. Retrieved from http://www.robocupjunior.org.au/acara

National Economic Council, Council of Economic Advisers, & Office of Science and Technology Policy. (2011). *Strategy for American innovation—Securing our economic growth and prosperity*. Retrieved from https://obamawhitehouse.archives.gov/sites/default/files/uploads/InnovationStrategy.pdf

Next Generation Science Standard. (2013). *Appendix I—Engineering design in the NGSS*. Retrieved from http://www.nextgenscience.org/sites/ngss/files/AppendixI-Engineering.Design.in.NGSS-FINAL_V2.pdf

Nourbakhsh, I. R., Hamner, E., Crowley, K., & Wilkinson, K. (2004). *Formal measures of learning in a secondary school mobile robotics course*. Paper presented at the 2004 IEEE International Conference on Robotics & Automation, New Orleans, LA.

OECD. (2015). *OECD innovation strategy 2015—An agenda for policy action*. Retrieved from Paris, France: http://www.oecd.org/sti/OECD-Innovation-Strategy-2015-CMIN2015-7.pdf

Oppliger, D. (2002, November). *Using FIRST LEGO league to enhance engineering education and to increase the pool of future engineering students (work in progress)*. Paper presented at the 32nd ASEE/IEEE Frontiers in Education Conference, Boston, MA.

OUTREACH@DARPA.MIL. (2014, March 13). *The DARPA grand challenge: Ten years later—Autonomous vehicle challenge led to new technologies and invigorated the prize challenge model of promoting innovation*. Retrieved from http://www.darpa.mil/news-events/2014-03-13

Papert, S. (1993). *Mindstorms—Children, computers, and powerful ideas* (2nd ed.). New York, NY: Basic Books.

Papert, S., & Harel, I. (1991). *Constructionism*. New York, NY: Ablex Publishing Corporation.

Partnership for 21st Century Skills. (2008). *21st Century skills, education & competitiveness guide—A resource and policy guide*. Retrieved from http://www.p21.org/storage/documents/21st_century_skills_education_and_competitiveness_guide.pdf

Piaget, J. (1929). *The child's conception of the world*. New York: Harcourt, Brace and Company.

Piaget, J. (1954). *The construction of reality in the child*. New York: Basic Books.

Robinson, K. (2010). *Changing education paradigms*. Retrieved from http://www.ted.com/talks/ken_robinson_changing_education_paradigms.html

Rogers, C., & Portsmore, M. (2004). Bringing engineering to elementary school. *Journal of STEM Education, 5*(3&4), 17–28.

Rusk, N., Resnick, M., Berg, R., & Pezalla-Granlund, M. (2008). New pathways into robotics: Strategies for broadening participation. *Journal of Science Education and Technology, 17*(1), 59–69.

Sklar, E., & Eguchi, A. (2004). RoboCupJunior—Four years later. In *Proceedings of RoboCup-2004: Robot Soccer World Cup VIII*.

Sklar, E., Eguchi, A., & Johnson, J. (2002). Examining the team robotics through RoboCupJunior. In *Proceedings of the Annual Conference of Japan Society for Educational Technology*.

Sklar, E., Eguchi, A., & Johnson, J. (2003). Scientific challenge award: RoboCupJunior—Learning with educational robotics. *AI Magazine, 24*(2), 43–46.

SoftBank Mobile Corp., & Aldebaran Robotics SAS. (2014). *SoftBank mobile and Aldebaran Unveil "Pepper"—the world's first personal robot that reads emotions*. Retrieved from http://www.softbank.jp/en/corp/group/sbm/news/press/2014/20140605_01/

Tanabe, K. (2015, June 23). Second generation Pepper for household use came out with a totally different "character" (Japanese). *Toyo Keizai.* Retrieved from http://toyokeizai.net/articles/-/74275

Tanenbaum, C. (2016). *STEM 2026: A vision for innovation in STEM education.* Retrieved from http://www.air.org/resource/stem-2026

The Conference Board of Canada. (n.a.). Innovation skills profile 2.0.

The White House. (2011). *Innovation.* Retrieved from http://www.whitehouse.gov/issues/economy/innovation

U.S. Department of Education. (2015). *Science, technology, engineering and math: Education for global leadership.* Retrieved from http://www.ed.gov/stem

U.S. Department of Education, & Office of Innovation and Improvement. (2016). *STEM 2026: A vision for innovation in STEM education.* Washington, DC: U.S. Department of Education, Office of Innovation and Improvement.

VEX EDR. (n.a.). *Standards matching & accreditation.* Retrieved from http://curriculum.vexrobotics.com/teacher-materials/standards-matching-and-accreditation

Wing, J. M. (2006). Computational thinking. *Communications of the ACM, 49*(3), 33–35.

Wing, J. M. (2010). *Computational thinking: What and why?* Retrieved from http://www.cs.cmu.edu/~CompThink/resources/TheLinkWing.pdf

Yamakami, T. (2012). *Innovation literacy: Implications from a shift toward dynamic multidisciplinary engineering.* Paper presented at the 8th International Conference on Information Science and Digital Content Technology (ICIDT), Jeju Island, Korea.

Chapter 2
Systems Thinking Approach to Robotics Curriculum in Schools

Christina Chalmers and Rod Nason

Abstract This chapter presents a systems thinking approach for the conceptualization, design, and implementation of robotics curriculum to scaffold students' learning of important Science, Technology, Engineering, and Mathematics (STEM) concepts and processes. This approach perceives the curriculum as a system of integrated elements and allows for the investigation of the interdependencies amongst the elements and the dynamics of the curriculum as a whole. Through this approach, we believe that students can be provided with robotics curriculum units that facilitate the learning of STEM "Big Ideas" of and about STEM. A STEM "Big Idea" is central to the understanding and application of STEM across a wide range of fields, one that links numerous STEM discipline understandings. Robotics is a rich context in which students can establish deep knowledge and robust understanding of STEM "Big Ideas". Curriculum units based on this systems thinking approach can do much to ensure that students engaged in robotics activities focus not only on the completion of robotics tasks but also on the social construction of integrated networks of authentic STEM knowledge centred around "Big Ideas" of and about STEM.

Keywords Robotics · STEM · Systems thinking · Curriculum

2.1 Introduction

In the process of designing and programming robots, students can learn many important Science, Technology, Engineering, and Mathematics (STEM) concepts and processes (Cejka et al. 2006). Unfortunately, this potential for advancing the learning of STEM through robotics is far from being realized. The challenge is to maintain student interest whilst not missing STEM "teaching moments" that allow

C. Chalmers (✉) · R. Nason
Queensland University of Technology, Brisbane, QLD, Australia
e-mail: c.chalmers@qut.edu.au

© Springer International Publishing AG 2017
M.S. Khine (ed.), *Robotics in STEM Education*,
DOI 10.1007/978-3-319-57786-9_2

students to go beyond trial-and-error strategies, behaviours that lead to weak solutions and limited learning (Barak and Zadok 2009).

To address this issue, we are proposing a systems thinking approach for the design and implementation of curriculum units with robotics. A systems thinking approach to curriculum focuses on the core areas of planning, design, implementation, and assessment and examines how these areas are aligned and connected to support students' success in learning (Jasparro 1998). Through this approach, we believe that students can be provided curriculum units to facilitate the learning of important STEM concepts and processes.

We begin by first discussing conceptions *of* and *about* STEM knowledge. This is because conceptions *of* and *about* STEM knowledge influence decisions teachers make about organizing learning experiences, teaching methodologies, and modes of assessment (Bencze et al. 2006). This viewpoint is borne out by a review of educational robotics literature from the last fifteen years. In most exemplary robotics units, there usually is an overt focus on the creation, testing, and advancement of knowledge *of* and *about* robots and the units are usually underpinned by systems thinking conceptions *of* and *about* STEM knowledge. Exemplary units are defined as units where the focus was on "doing with understanding"; that is, the focus of the units went beyond the mere completion of robotics task(s) and extended to the social construction of knowledge *of* and *about* STEM. This is characterized in most cases by an emphasis on STEM "Big Ideas".

A "Big Idea" is a statement of an idea that is central to the understanding and application of STEM across a wide range of fields, one that links numerous understandings into coherent wholes (Charles 2005). By encompassing and connecting concepts, "Big Ideas" *of* STEM can provide an organizing structure for content knowledge about robotics (Silk 2011) and form a basis for facilitating the meaningful learning of STEM knowledge. A sample of STEM concepts is presented in Table 2.1 that have applications within the contexts of robotics activities. The connection of concepts and ideas can help to establish strong conceptual links within and between the STEM disciplines (c.f., Charles 2005; Harlen 2010).

Focusing on "Big Ideas" can facilitate the meaningful learning of STEM knowledge in robotics contexts by helping students connect concepts and showing students how STEM knowledge can provide them with ways of thinking about and making sense of the world (Harlen 2010; Lesh and Doerr 2003). For example, *Proportional Reasoning is central to how the Motion of the robot can be controlled through Programming*, as the relationships between the construction of the robot, the values used to program the robot, and the movement of the robot are often proportional in nature (Silk 2011). Thus, the exploration of proportional reasoning within the context of robotics can enable students to better understand the application of proportional reasoning in their everyday worlds. Furthermore, the exploration of "Big Ideas" can enable students to progress beyond trial-and-error problem-solving strategies in robotics and many other STEM learning activities.

Table 2.1 A sample of STEM understandings within the context of robotics

	Ideas of STEM	Ideas about STEM
Science	Energy (Rockland et al. 2010), Force (Cejka et al. 2006; Chambers et al. 2007; Rockland et al. 2010) Motion (Chambers et al. 2007; Williams et al. 2007)	Inquiry (Rockland et al. 2010; Sullivan 2008; Williams et al. 2007) Process (Sullivan 2008)
Technology	Design (Sullivan 2008) Programming (Cejka et al. 2006; Silk 2011; Sullivan and Heffernan 2016) Systems (Grubbs 2013; Sullivan 2008)	Computational thinking (Berland and Wilensky 2015; Bers et al. 2014; Sullivan and Heffernan 2016), Systems thinking (Berland and Wilensky 2015)
Engineering	Structures (Cejka et al. 2006) Simple machines (Gears, Levers, Pulleys) (Cejka et al. 2006; Chambers et al. 2008; Williams et al. 2007)	Design process (Bers and Portsmore 2005; Cejka et al. 2006; Grubbs 2013; Rockland et al. 2010)
Mathematics	Proportional reasoning (Silk et al. 2010) Ratio (Silk 2011) Distance, Measurement (Grubbs 2013)	Problem-solving (Chalmers 2013; Norton et al. 2007; Sullivan and Heffernan 2016)

Therefore, we are proposing that curriculum units with robotics should be based on a systems thinking viewpoint about STEM knowledge and focus on the construction of "Big Ideas" *of* and *about* STEM. The implementation of curriculum units with robotics centred on "Big Ideas" can do much to ensure that students engaged in robotics activities not only focus on the satisfactory completion of robots but also on construction of authentic knowledge *of* and *about* STEM. This clearly has implications not only for aims and objectives but also for other key elements in the process of developing curriculum units with robotics such as follows:

- Framing robotics learning activities;
- Integrating robotics learning activities into STEM curriculum units;
- Selection and utilization of thinking tools; and
- Design and implementation of assessment.

Each of these four elements will now be discussed in turn during the following sections of this chapter. We conclude this chapter by integrating these elements into a systems framework to facilitate the design of curriculum units with robotics to scaffold the learning of STEM "Big Ideas".

2.2   Framing Robotics Learning Activities

A review of the literature from the fields of *model-eliciting activities* (MEAs), *learning-from-design, educational robotics*, and *knowledge-building* indicates that how STEM learning activities are framed influences whether or not they facilitate the construction of STEM "Big Ideas". Therefore, in this section, a system of six principles for framing robotics learning activities to facilitate the construction of STEM "Big Ideas" is presented (see Fig. 2.1).

1. *Foregrounding Principle: A robotics learning activity should focus on targeted STEM concepts that are repeatedly foregrounded during the course of the activity.*

   This principle has its genesis in findings from the Robots Algebra Project (Silk et al. 2010), the Programmable Bricks Project (Rusk et al. 2008), Tufts University's Center for Engineering Education and Outreach research (Rogers 2012; Wendell and Rogers 2013), and research conducted by the Robotics@QUT program (Chalmers 2013). A clear outcome from this research is that the content of robotics activities needs to be targeted and precise. This can be achieved by repeatedly "foregrounding" the targeted "Big Idea(s)" during the course of an activity (Silk et al. 2010) and by highlighting particular ideas and concepts in the natural course of working on a project (Rusk et al. 2008).

2. *Sustained Knowledge-building Principle: A robotics learning activity should be meaningful and relevant to students and motivate students to make sense of the situation based on the extension of their personal knowledge and experiences.*

Fig. 2.1 System of principles for framing robotics learning activities

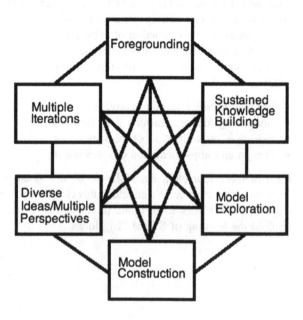

This principle was derived from educational robotics research (e.g. Bers 2008; Bers et al. 2014; Rusk et al. 2008) and MEAs research (e.g. Lesh and Doerr 2003; Hamilton et al. 2008; Yildirim et al. 2010). Different students are attracted to different types of robotics activities (Bers 2008). This clearly implies that teachers need to not only carefully select contexts that can stimulate the interest and involvement of a diverse student population; they also should consider thinking beyond traditional technological approaches when engaged in the process of framing a robotics activity (Rusk et al. 2008). Non-technological approaches such as the social narrative approach (Hamner et al. 2008), the arts and engineering approach (Rusk et al. 2008), the literature and robotics approach (Bers 2008), and the robotics and emotional competency approach (Bers et al. 2014) where robotics becomes a tool rather than the focus of the activity can provide the means for broadening active participation by students with non-technological interests. Motivating active participation in an activity is a necessary but not sufficient condition for sustained knowledge-building. MEAs research has found that to establish and maintain knowledge-building engagement, STEM learning activities also need to motivate students to make sense of the problem context by extending on their personal knowledge and experiences (Hamilton et al. 2008; Yildirim et al. 2010).

3. *Diverse Ideas/Multiple Perspectives Principle: A robotics learning activity should put students in situations where diverse ideas and/or alternative perspectives can emerge and be juxtaposed.*

Research findings from the fields of MEAs (e.g. Lesh et al. 2003) and knowledge-building communities (e.g. Scardamalia 2002) clearly indicate that the development of "Big Ideas" is facilitated if a diversity of ideas and/or multiple perspectives are brought to a problem. MEA research has found that closely juxtaposing multiple perspectives helps students overcome conceptual egocentrism and centring that are especially apparent when unstable conceptual systems are used to make sense of experiences (Lesh et al. 2003). The juxtaposing of different perspectives can help shift focus to the big picture and encourage students to generalize patterns and relationships. Therefore, integrating multiple perspectives encourages students to think more deeply about their experiences.

This principle can be enacted by the following:

- Formation of teams consisting of members with different technical capabilities, different cognitive styles, or different prior experiences (Lesh et al. 2003; Scardamalia 2002);
- Encouraging members of teams to play different roles such as manager, monitor, recorder, data gatherer, or tool operator (Chalmers 2009; Lesh et al. 2003);
- Having students serve as editorial boards that assess strengths and weaknesses of other teams' proposals or by introducing the role of a client (Lesh et al. 2003);

- Utilization of reflection tools to think about group functioning, about roles played by different individuals, and about ideas and strategies that were and were not productive (Chalmers and Nason 2005; Lesh et al. 2003).

4. *Model Construction Principle: A robotics learning activity should create the need for problem resolution via the construction/modification of a model that is powerful (in the specific situation), sharable (with others), easily modified, and reusable (in other situations).*

Research from the field of MEAs (e.g. Lesh and Doerr 2003; Hamilton et al. 2008; Yildirim et al. 2010) and from the Robots Algebra Project (Silk 2011; Silk et al. 2010) has found that requiring students to construct and/or modify a model that is capable of being used by others in similar situations, and robust enough to be used as a tool in other STEM learning can significantly increase the probability that the construction/advancement of "Big Ideas" *of* and *about* STEM will occur. In robotics activities, a model could be a generalizable procedure for designing/constructing a robot, a flowchart, a "how-to" toolkit, a set of rules and/or specifications, a prediction model, a judging scheme, a method, an index, or a metaphor for seeing or interpreting things (Silk et al. 2010).

5. *Model Explanation Principle: A robotics learning activity should require students to explicitly reveal how they generated a model.*

Findings from MEAs, learning-from-design, and educational robotics research indicate that construction of "Big Ideas" *of* and *about* STEM is facilitated if learning activities are thought-revealing in nature (Hamilton et al. 2008; Sadler et al. 2000; Silk et al. 2010; Yildirim et al. 2010). That is, they require students to reveal not only their models but also the thoughts underlying the development of their models. According to Sadler et al. (2000), students' thought-revealing explanations elicited by the framing of the learning activity should be formative, capturing all attempts and trials. This enables students to examine their progress, assess the evolution of the model, and reflect about the model (Yildirim et al. 2010). With MEAs, model explanation typically involves students writing memo(s) to a client describing their model and documenting how it was developed (Lesh and Doerr 2003). Engineering design research journals and diaries have also been successfully utilized by learning-from-design studies (e.g. Wendell and Kolodner 2014), and in educational robotics research (e.g. Bers et al. 2014; Hamner et al. 2008), to generate thought-revealing explanations by students.

6. *Multiple Iterations Principle: A robotics leaning activity should require students to plan and make multiple iterations to not only to their robotic design but also iterative refinements in understanding of STEM concepts and processes.*

Taking the time to plan can improve both design product and learning outcomes (Fortus et al. 2004). Unfortunately, students often do not engage in the planning phase of the design process (Rogers and Wallace 2000; Welch 1999). MEA researchers such as Hamilton et al. (2008) have found that this can be addressed by requiring students to present their initial plans (and iterative revisions of these plans) in their reports to clients. This dilemma also can be addressed by constraining the time and resources available for a learning activity; if students are given too much time and too many materials, they often resort to trial-and-error methods rather than advance planning to complete many design problems (Wendell and Kolodner 2014). Ensuring that the artefacts (i.e. robots) to be designed are relatively easy to build and/or modify has been found to facilitate multiple iterations and testings of ideas (Sadler et al. 2000).

Research from the field of learning-from-design indicates that designing a working artefact involves iterative design. Iterative design of an artefact affords opportunities for students to incrementally construct, evaluate, discuss, and revise both the models they are designing and their conceptions (Puntambekar and Kolodner 2005). Iterative design also enables students to learn from their failures as well as their successes (Sadler et al. 2000). However, this iterative process can lead to significant student frustration and discouragement if students do not shift from the typical trial-and-error design process (Bers et al. 2002) and use STEM "Big Ideas" to predict what their robot will do and to choose the best solution (Silk 2011).

2.2.1 Application of the System of Principles

Robotics activities are engaging, however, how the activities are framed influences whether or not they facilitate the construction of STEM "Big Ideas". As is indicated in Fig. 2.1, the six principles should be conceptualized not as a sequential list but as a system of interrelated principles that are implemented iteratively in cycles with multiple feedback loops. When implemented as a system, the six principles can be utilized to guide not only the design of new robotics activities but also the evaluation and modification of existing robotics activities to ensure that they facilitate the learning of "Big Ideas" of and about STEM. The principles can help develop curriculum units that focus attention on the "Big Ideas", promote active inquiry, and support students reflecting on the learning process.

2.3 Integrating Robotics Learning Activities into STEM Curriculum Units

Isolated problem-solving activities such as robotic design tasks are seldom enough by themselves to ensure the learning of STEM "Big Ideas" (Chambers et al. 2008; Silk 2011). Sequences of structurally related STEM learning activities conducted over a number of class periods in conjunction with discussions and explorations focusing on structural similarities amongst the related activities also are needed

Fig. 2.2 System of modules
for developing sequences of
structurally related STEM
learning activities

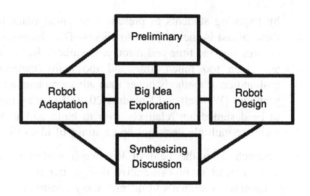

(Lesh et al. 2003). Therefore, in this section a system of five modules for developing sequences of structurally related STEM learning activities in curriculum units with robotics is presented (see Fig. 2.2). The system is derived from an analysis and synthesis of research on the design of structurally related activities from the fields of MEAs, learning-from-design, and educational robotics.

2.3.1 Preliminary Activity

The Preliminary Activities familiarize students with the context of a robotics construction task. Familiarizing students with the context of a task can be achieved by providing students with background information via short articles, webpages, and video clips accompanied by questions that:

- Familiarize students with the context of the robotics design problem so that their solutions are based on extensions of students' real life knowledge and experiences; and
- Build up "minimum prerequisites" for students to begin working on the robotics design problem (Lesh and Doerr 2003; Silk 2011).

2.3.2 Robot Design Activity

During the course of each class period when the students are engaged in the process of designing and constructing a robot, teachers can do much to facilitate a culture of metacognition and knowledge-building by asking students questions that require them to:

- Predict outcomes—This can help students to understand what kinds of information they might need to successfully complete the robotic design task (Darling-Hammond et al. 2008; Puntambekar and Kolodner 2005); and
- Monitor what they are doing—For example, "What are you working on now?", "Why are you working on it?", and "How does it help you?" (Darling-Hammond et al. 2008; Puntambekar and Kolodner 2005).

At the end of each class period during the course of a Robot Design Activity, the knowledge-building aspects of the activity can be further enhanced by the utilization of Pin-up Sessions, Presentations and Discussions, and Reflection and Debriefing Activities.

In Pin-up Sessions, students periodically present their design ideas and sketches by creating a poster, pinning it to the wall, and then explaining to the class their intentions and how they plan to achieve them (Puntambekar and Kolodner 2005). The primary goals here are to encourage students to think through their ideas deeply and make their reasoning clear. According to Puntambekar and Kolodner, hearing the ideas of others provides grist for students to learn what makes for good justifications. They also argue that Pin-up Sessions give an entire class a chance to consider additional alternatives besides the ones they had considered in their small groups. Additionally, Pin-up Sessions may give groups that are experiencing difficulties a chance to come up to the level of the rest of the class.

Presentations and Discussions are whole-class activities in which students make formal presentations about what they have created and how they created it. The primary goals here are for students to explain their work, see other students' alternative approaches and outcomes, discuss strengths and weaknesses, and identify directions for improvement for their own work and the work of others. Puntambekar and Kolodner (2005) suggest that Presentations and Discussions should occur between iterations of the design process to provide students with opportunities for talking about STEM ideas and seeing how others are applying them. Because they enable students to engage in discourse about ideas, critique in a constructive way, and revise work after receiving feedback, Presentations and Discussions can do much to facilitate the development of metacognitive thinking skills such as reflecting on and regulating learning (Bers et al. 2002; Darling-Hammond et al. 2008). Lesh et al. (2003) have found that having teams of students produce executive summaries is a most effective means for facilitating knowledge-building Presentations and Discussions and the development of metacognitive thinking. Other means suggested by Lesh et al. (2003) for facilitating knowledge-building Presentations and Discussions and the development of metacognitive thinking include multimedia presentations and having teams of students play the role of clients who give feedback to other teams about the strengths and limitations of their models.

Reflection and Debriefing Activities' primary goal is to help students assume a reflective and strategic stance towards learning (Darling-Hammond et al. 2008) and adopt an "increasingly productive personae for learning and problem solving (Lesh et al. 2003, p. 50)". This process can be much facilitated by the utilization of reflection tools that focus not only on solutions, but also on group dynamics and the roles that individual students played during different stages of the solution process (Chalmers 2009).

In order to ensure that student teams do not "give up" when experiencing frustration and/or failure, they should be provided with adequate "self-help" resources such as "*just-in-time*" "*how-to*" *toolkits* (Lesh and Doerr 2003). Examples of such toolkits that could be utilized in Robot Design Activities are online tutorials, simple building and programming instructions, tutorials, video tutorials, and manuals.

2.3.3 *"Big Idea" Exploration Activity*

The primary goal here is to form a cognitive link between robotics and non-robotics contexts of the "Big Idea(s)" foregrounded in the robotics tasks. Research in the field of robotics in schools (e.g. Chambers et al. 2008) suggests that providing students with physical experiences such as designing robots is not enough by itself for students to develop understandings of STEM "Big Ideas". As Lesh et al. (2003) point out, to help students go beyond thinking *with* a "Big Idea" and also think *about* it, several structurally similar embodiments are needed. Students also need to focus on similarities and differences as the relevant "Big Idea(s)" function in different contexts. Thus, students must go beyond investigating individual ideas to investigate structure-related relationships amongst several alternative embodiments —perhaps by making translations or predictions from one context to another.

2.3.4 *Robot Adaptation Activity*

The primary goal here are to have students deal with robotics problem(s) similar to but more complex than those addressed in a Robot Design Activity and whilst in the process have them adapt and/or extend the "Big Idea(s)" developed and refined in the Robot Design and/or the "Big Idea" Exploration Activities. A good example of a Robot Adaptation Activity is provided by the Robot Synchronized Dancing Activity created by Silk (2011). In this activity, students were required to create a "how-to" toolkit to coordinate the physical features, program parameters, and robot movements of many different existing robots so that they can "dance in sync" with each other.

2.3.5 Synthesizing Discussion

Synthesizing Discussions are conducted during the concluding phase of a sequence of learning activities (Lesh et al. 2003). These discussions provide closure and have students go beyond thinking with the foregrounded "Big Idea(s)" and advance towards making the "Big Idea(s)" explicit knowledge objects of thought that serve in the further advancement of knowledge. Knowledge-building and MEA research indicate that this process can be facilitated by whole-class teacher-led activities that focus on structural similarities and differences between the different embodiments of the "Big Idea(s)" explored during the course of the sequence of learning activities.

2.3.6 Application of the System of Modules

Because they are conceptualized as a system, the modules can be sequenced in different ways to facilitate the:

- Introduction of "Big Idea(s)" through robotics;
- Application and Extension of "Big Idea(s)" through robotics; and
- Introduction, Application, and Extension of "Big Idea(s)" through robotics.

Thus, the sequence in which the modules are applied to structure a curriculum unit with robotics can be varied to meet the needs/preferences of teachers/researchers. For example, if a teacher/researcher wanted to introduce the STEM ideas of rotational speed and torque through robotics, the unit could begin with a Preliminary Activity where students are presented with a real-world situation establishing the need for the development of a set of specifications (i.e. a model) for the design of a rescue vehicle capable of travelling across flat surfaces quickly, but also able to negotiate steep hills. During the process of developing and refining the model during the Robot Design Activity, the teams of students could investigate what gear trains yield the best compromise between rotational speed and torque and develop key science/engineering ideas such as follows:

- Gearing down will slow down your robot but will supply more power for climbing; and
- Gearing up will speed up your robot but will supply less power for climbing.

In order to extend the understandings of rotational speed and torque developed during the course of the Robot Design Activity, these understandings could be further developed in non-robotics contexts such as bicycle gears during the course of the "Big Idea" Exploration Activity. They also could be extended in the Synthesizing Discussions where the notion that gears work on the principle of mechanical advantage can be explored and generalized across different contexts.

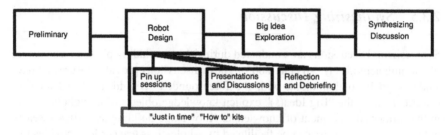

Fig. 2.3 Introduction of "Big Idea(s)" through robotics

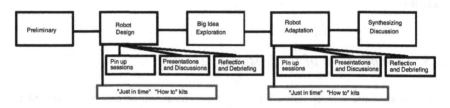

Fig. 2.4 Introduction, Application, and Extension of "Big Idea(s)" through robotics

One application of the system of modules for structuring learning activities is encapsulated in Fig. 2.3. In this structuring of the modules, "Big Idea(s)" are introduced in the context of robotics design tasks. The "Big Idea(s)" foregrounded in the robotic learning activity are further explored and extended during the course of the "Big Idea" Exploration Activity and Synthesizing Discussions.

The application of the system of modules encapsulated in Fig. 2.4 is similar in most respects to that encapsulated in Fig. 2.3. The major difference is the inclusion of the Robot Adaptation Activity where students deal with robotics problem(s) similar to but more complex than those addressed in a Robot Design Activity.

Other different ways for utilizing the system of modules to apply and extend "Big Idea(s)" through robotics are illustrated in Fig. 2.5. In this structuring, the initial preliminary or "Big Idea(s)" activities are explored first and then students are required to extrapolate and apply their understandings of STEM "Big Idea(s)" foregrounded in these activities to scaffold the design, construction, and programming of their robot during Robot Design and/or Robot Adaptation Activities.

As was noted earlier, the sequence in which the modules for developing sequences are utilized to structure a curriculum unit with robotics with a focus on STEM "Big Idea(s)" Exploration can be varied to meet the needs/preferences of teachers (and researchers). However, whilst in the process of framing robotic learning activities and integrating them into a sequence of structurally related STEM learning activities, teachers (and researchers) also need to concurrently consider how thinking tools can be utilized during the course of the curriculum unit with robotics to further scaffold the learning of STEM "Big Ideas".

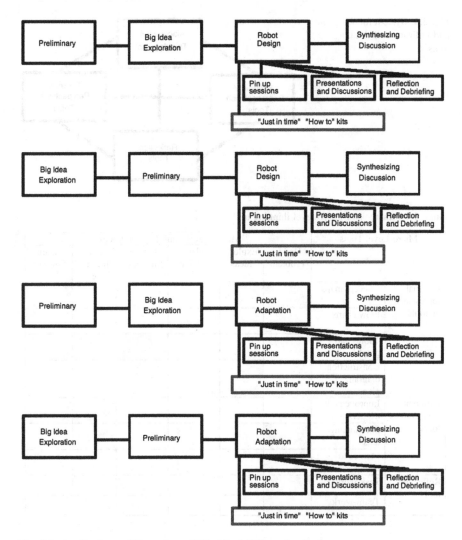

Fig. 2.5 Application and Extension of "Big Idea(s)" through robotics

2.4 Selection and Utilization of Thinking Tools

Thinking tools have important roles in supporting the learning of "Big Ideas" *of* and *about* STEM during the course of design activities (Kokotovich 2008; Puntambekar and Kolodner 2005). Therefore, decisions about what thinking tools can be utilized and how they should be utilized during the course of a curriculum unit with robotics need much thought. To facilitate this process, in this section we present a system for the selection and utilization of thinking tools (see Fig. 2.6).

Fig. 2.6 System for selection and utilization of thinking tools

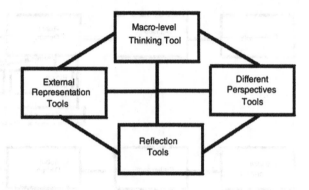

Table 2.2 Macro- and micro-level thinking tools

Micro-Level Tools		Macro-Level Tool: Design Process					
		Define Problem	Explore Ideas	Plan/ Design	Create/ Implement	Test/ Improve	Share Solution
External representation tools	Concept maps	■	■				■
	Flow charts			■			■
	Tables and graphs				■		■
	Construction diagrams/ plans			■	■		
Different perspectives tools	Improvement triggers			■	■	■	
	Six thinking hats	■		■			■
	Memos to clients				■		■
Reflection tools	Task-work	■			■	■	
	Team-work	■			■	■	

Research literature from the fields of *learning-from-design* and *educational robotics* indicates that both macro- and micro-level thinking tools are needed to facilitate learning. Thus, our proposed system consists of three types of micro-thinking tools (external representation tools, different perspective tools, and reflection tools) integrated into the operation of a macro-level tool (see Table 2.2).

2.4.1 Macro-level Thinking Tools

A macro-level thinking tool (e.g. the Engineering Design Process) provides students with a global framework to guide them through the major steps of the design process whilst also enabling them to go back when necessary to earlier steps to make modifications or changes to a design (Puntambekar and Kolodner 2005). The ultimate goal of these macro-level tools is to help students to create the best design possible by improving it over and over again. These tools are not linear but cyclical in nature. This means that each of the steps in these tools may be repeated as many times as needed, making improvements along the way. For example, after testing a design and finding a problem, the macro-level tools allow you to go back to an earlier step to make a modification or change to a design.

2.4.2 Micro-level Thinking Tools

Within the design process, micro-level tools usually have three main roles:

- Generating external representations;
- Looking at a design problem from different perspectives; and
- Promoting reflection.

2.4.3 External Representation Tools

External representation tools (such as those listed in Table 2.2) facilitate the construction of external representations that help learners to collect, organize, absorb, and understand information, advance knowledge (Caviglioli et al. 2002), and make sense of messy situations inherent within many design tasks (Fathulla and Basden 2007). During the early steps of the design process, external representations mediated by these tools help learners to structure and map the salient issues, thoughts, and ideas relevant to a design problem (Kokotovich 2008). The construction of the external representations during the early steps of the design process also helps students to think logically about the problem and define it in a more holistic way, rather than just jumping in and relying on trial-and-error strategies (Kokotovich 2008; Norton et al. 2007). This places students in a position to develop more considered responses to the design problem.

The external representations mediated by these tools also enable learners to identify and externalize their models of understanding (Caviglioli et al. 2002) and make their thinking visible during the middle and later phases of the design process (Lane 2013). By making their thinking visible, learners are able to further analyse their understanding of a design problem and add to, adapt, and change particular aspects of their model of understanding (Caviglioli et al. 2002). As Lane (2013) points out, the rigour of creating an external representation such as a diagram, iterating it through several revisions and of clearly specifying the purposes and assumptions behind it is enough to change an individual's thinking about a design problem just as writing text helps organize and present one's thoughts to others.

Thus, these tools have the potential to facilitate learning throughout the whole design process. This places students in a position to develop more considered responses to the design problem. The external representation generated by these tools can play important roles in facilitating planning how-to approach a design problem, monitoring progress towards the solution to the problem, and evaluating progress towards the completion of the task (Norton et al. 2007).

These thinking tools can also facilitate the advancement of learning and understanding at the group level during all steps of the design process (Lane 2013). The creation of external representations during early steps of the design process enables each learner to share his/her thinking or understanding of a design problem with other students and mediate the development of shared knowledge and understanding about the problem. In order to develop a shared understanding and knowledge of a problem, group members must first negotiate a shared external representation or model of the problem (Chalmers 2009; Fiore and Schooler 2004). Shared external representations facilitate the process of articulating students' thinking and allow group members to formulate an accurate shared understanding (Cannon-Bowers and Salas 2001).

During the middle and later steps of the design process, a shared external representation can provide a focus for discussion and contribute to the collective manipulation, reconstruction, and reinterpretation of information and ideas (Lane 2013). The collective manipulation, reconstruction, and reinterpretation of information and ideas mediated by iterative modifications of shared external representations scaffold the advancement of understanding and knowledge by the group (Scardamalia 2002).

2.4.4 Different Perspective Tools

When teams of students engaged in design tasks seem entrenched in a particularly unproductive mindset (e.g. focusing on trial-and-error strategies) and/or are confronted by impasses (e.g. finding that the artefact they are creating does not work as well as expected), they more often than not are not utilizing metacognitive thinking and systematically reasoning about what they could do to move forward (Puntambekar and Kolodner 2005). This can be addressed by the utilization of thinking tools such as Improvement Triggers (Eberle 1997) that enable students to look at the design problem from different perspectives. Improvement Triggers are a list of SCAMPER (Substitute, Combine, Adapt, Modify, Put to another use, Eliminate, Reverse) questions that can help students look at their work from different viewpoints. These questions can focus on robot design/construction issues (e.g. Adapt: What ideas could we use to adapt or readjust to improve our robots? What other ideas could we use for inspiration for our robots?) and/or the relationships between STEM "Big Ideas" and robotics (e.g. Eliminate: How could we streamline or simplify our models? What elements of our models could we remove?). Other tools that can be used for this purpose are: Six Thinking Hats (de Bono 1985) and Memos to Clients (Lesh and Clarke 2000).

2.4.5 Reflection Tools

Reflection tools have important roles in promoting reflection about task-work and team-work prior, during and after the completion of design problem activities (Chalmers 2009; Hamilton et al. 2008). Reflection tools help students recall and then record significant aspects about what they have done and thus enable students to: (a) relate new knowledge to their prior understanding, (b) mindfully abstract knowledge, and (c) understand how their learning and problem-solving strategies might be reapplied (Hmelo-Silver 2004).

Many external representation and different perspective tools can be utilized to promote reflection about task-work. Reflection about task-work also can be facilitated by sets of reflection questions that students are required to answer following a robotic activity. These questions can be the focus of the class discussions that follow the activity (Hamilton et al. 2008). A review of the literature (e.g. Hamilton et al. 2008; Lesh et al. 2003; Silk 2011) indicates that reflection questions should focus not only on robot design and construction but also on relationships between the robotics activity and STEM concepts and processes.

Research literature from the fields of MEAs (e.g. Hamilton et al. 2008), collaborative learning (e.g. Barron 2000), cooperative learning (e.g. Johnson and Johnson 2004), and team-work (Beatty and Barker 2004) indicates that reflection on team-work can also be facilitated by tools that attune students to:

- Individual roles (e.g. How did your individual roles change during the course of the design process and why?);
- Organization of group-work (e.g. How did you organize your group-work? What strategies did your group use to develop new ideas, interpretations or hunches?);
- Monitoring and improvement of team-work (e.g. How were good ideas shared within your group? What are two things your group is doing well and one thing that needs to improve? How did you monitor the effectiveness of your group-work? What could you do to improve the effectiveness of your group?);
- Problems encountered and how they were resolved (e.g. What problems did you encounter in working as a group and how did you resolve them?); and
- Planning for the future (e.g. If you were to embark on a second, similar task as a group, what would be different about the way you go about working, and why?).

2.4.6 Application of the System of Thinking Tools

A review of the literature indicates that for optimal impact on student learning, thinking tools need to operate in a synergic manner to:

- Help students recognize what step of the design process they are in and to record ideas and knowledge relevant to that step;
- Provide prompts and explanations to help students decide how-to move forward during each step of the design process;
- Provide guidance for students both in carrying out design activities and reflecting on them in order to learn from them; and
- Encourage students to think about and articulate what they have done and why without diverting too much of their time from the raison d'etre of a curriculum unit, the construction of robotic artefacts and STEM "Big Idea" knowledge artefacts (Bers et al. 2002; Lesh and Clarke 2000; Puntambekar and Kolodner 2005).

The clear implication of this is that the number of micro-level tools subsumed within the operation of the macro-level tool needs to be limited. However, this is not necessarily a problem; each micro-level tool can be utilized more than once during the design process (see Table 2.2). Indeed if maximum impact is desired, then each selected micro-level tool should be utilized more than once during the course of the curriculum unit (Lesh and Clarke 2000; Puntambekar and Kolodner 2005). Therefore, limiting the number of micro-level tools utilized within a curriculum unit with robotics probably has positive rather than adverse effects on student learning, especially if each of the selected micro-level tools is utilized in multiple steps of the design process.

2.5 Design and Implementation of Assessment

Because assessment sends a clear message to students about what is worth learning, how it should be learned, and how well we expect them to perform it is imperative that assessment be philosophically consistent with the pedagogical framework implicit in the learning activities. This system is consistent with the constructionist framework (Papert 1980) implicit in previous sections of this article. The system is derived from an analysis and synthesis of the literature from the fields *model-eliciting activities, learning-from-design, educational robotics,* and *assessment theory*. Therefore, in this section we present a system for the design and implementation of assessment in curriculum units with robotics (see Fig. 2.7). Focussing on both summative and formative assessment this system of gives teachers opportunities to assess the learning process as well as the end product.

As is indicated in Table 2.3, our system has four categories of artefacts. The four categories identified address both formative and summative assessment. Category A

Fig. 2.7 System for the design and implementation of assessment

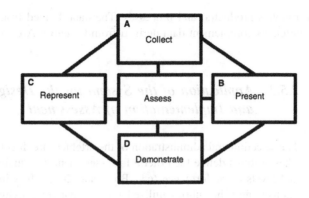

Table 2.3 Assessment artefacts

Category	Purpose	Assessment Artefacts
A. Collect	Collection of work selected to document progress within a given task	Portfolios[a]
		Engineering design notebooks[a]
		Robot design journals[a]
B. Present	Presentation of prototype(s), description of design solution and process, and description of rationale for arriving at their solution	Presentations[a]
		Demonstrations[a]
		Reports/Memos
		Poster sessions[a]
		Video journals[a]
		Exhibitions[a]
C. Represent	Representation of students' understanding of STEM concepts/processes	Representations produced with Representation Generating Tools[a] (see Table 2.2)
D. Demonstrate	Demonstration of students' understanding of STEM concepts/processes and robotics product (s) and processes	Observations[a]
		Interviews[a]
		Exams
		Reflective essays
		Robot challenge

[a]Change over time in the level of sophistication and complexity assessed

artefacts operate at the macro-level throughout the course of a curriculum unit and consist of a collection of student's work selected to document progress within a given task. Category A collections integrate data derived from Category B and C assessment artefacts. Category B artefacts consist of presentations of prototypes, descriptions of design solution and process, and justification of how students arrived at a solution. Category C artefacts focus on the representation of students' understanding of STEM concepts/processes. Finally, Category D artefacts demonstrate students' understanding of STEM concepts/processes and their robotics product(s) and processes. The data derived from Category D assessment artefacts complement data derived from Category A-C artefacts.

2.5.1 Application of the System for the Design and Implementation of Assessment

The selection and administration of the artefacts are directed by assessment rubrics. These rubrics define the criteria for assessment, the qualities that will be assessed, and levels of performance (c.f., Brookhart 2013). It is important that the selected artefacts and the rubrics utilized for summative assessment focus on assessment *about* learning and during formative assessment focus on assessment *for* learning (Black et al. 2008; Caitlin 2012). Through assessment for learning, teachers can ascertain students' knowledge, perceptions, and misconceptions and use this information to diagnose students' needs, provide them with constructive feedback, and plan interventions to support students to operate at the edge of their competence. Caitlin (2012) identified three essential elements of assessment for learning: Learning Intentions and Success Criteria, Quality Interactions and Feedback, and Peer Assessment. Together, these elements can provide students with prompts they can use to improve their quality of work, helping them feel in control of their learning (Stiggins et al. 2007), shaping, and improving their competence by short-circuiting the randomness and inefficiency of trial-and-error learning (Sadler 1989).

Learning Intentions are not learning aims or objectives but instead a student perspective; it also is about what students will learn, not what they will do (Caitlin, 2012). Caitlin suggests that where possible, the *Success Criteria* in curriculum units with robotics should focus on demonstration and process explanation. Teachers should establish Learning Intentions and Success Criteria by negotiation and students should record these on a pin-up board. Making them visible acts as a reference point throughout the activity helps keep students on task.

Quality Interactions and Feedback generally come in the form of teacher comments and/or guiding questions that encourage students to express and share ideas. Quality Feedback has to "strike a balance between students recognising what is good about their work, as well as what is necessary to improve" (Caitlin 2012, p. 7).

Peer Assessment can also assist students monitoring their learning and they can use the feedback from this monitoring to make adaptations and adjustments to what they understand (Earl 2003). The focus on student reflection is powerful in building metacognition and an ability to plan for future learning goal.

Whilst formative assessment needs to be consistent with the constructionist pedagogical framework underlying the learning activities (Wiggins and McTighe 2005). Summative assessment should also focus on determining to what extent the instructional/learning goals of the unit have been met (Stiggins et al. 2007). Adapting and applying the artefacts utilized in formative assessment for use in the summative assessment at the end of a curriculum unit can facilitate this. This integrates assessment into the teaching/learning process and encourages the active involvement of students in their learning (Earl 2003).

2.6 A Systems Framework to Facilitate the Design of Curriculum Units with Robotics

In this section, the four systems presented in the previous sections are integrated into a systems framework to facilitate the design of curriculum units with robotics that scaffold the learning of important STEM "Big Ideas" (see Fig. 2.8). At the core of this framework are STEM "Big Ideas".

There are many equally appropriate potential pathways in which the framework could be applied to facilitate the process of designing a curriculum unit with robotics. For example, the framework could enable teachers/researchers to begin the process by designing in order: the robotics learning activities, the sequence of structurally related STEM learning activities, the thinking tools, and finally the assessment artefacts and their associated rubrics. On the other hand, teachers/

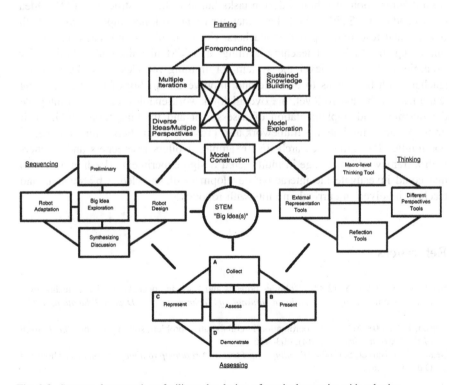

Fig. 2.8 Systems framework to facilitate the design of curriculum units with robotics

researchers also could utilize the framework to engage in "backward design" (Wiggins and McTighe 2005) and first work on the development of the assessment artefacts and their associated rubrics and have the assessment serve as a guide for directing the design of appropriate learning and thinking tools.

However, whatever pathway is utilized, it is important to note that the framework requires that teachers/researchers to conceptualize that:

1. The design of a curriculum unit with robotics is an iterative process (i.e. each element of the framework is revisited on multiple occasions during the design of the curriculum unit); and
2. STEM "Big Idea(s)" are at the core of the framework: therefore, constant recourse to them needs to be made during the design of a curriculum unit with robotics.

2.7 Concluding Remarks

This chapter has presented a systems framework to facilitate the design and implementation of curriculum units with robotics that scaffold not only the successful completion of robotics design tasks but also the construction of "Big Idea (s)" *of* and *about* STEM. We believe that our framework has implications for both practice and research. It provides teachers with both micro- and macro-means for improving the quality of teaching/learning of STEM in units with robotics. For example, at the micro-level the system for framing robotics activities provides teachers with the means to evaluate and improve the quality of robotics learning activities. At the macro-level, the overall framework enables teachers to integrate the planning and implementation of assessment and thinking tools within their units. At the same time, the framework also provides researchers with a number of possibilities for further research. For example, it offers researchers and teachers with a framework to engage in multi-tiered design experiments (Lesh et al. 2008) that could investigate the interactive development of knowledge by students and teachers involved in curriculum units with robotics.

References

Barak, M., & Zadok, Y. (2009). Robotics projects and learning concepts in science, technology, and problem solving. *International Journal of Technology and Design Education, 19*(3), 289–307.

Barron, B. (2000). Achieving coordination in collaborative problem-solving groups. *The Journal of the Learning Sciences, 9*(4), 403–436.

Beatty, C., & Barker, S. (2004). *Building smart teams: A roadmap to high performance.* Thousand Oaks, CA: Sage.

Bencze, J. L., Bowen, G. M., & Alsop, S. (2006). Teachers' tendencies to promote student-led science projects: Associations with their views about science. *Science Education, 90*(3), 400–419.

Berland, M., & Wilensky, U. (2015). Comparing virtual and physical robotics environments for supporting complex systems and computational thinking. *Journal of Science Education and Technology, 24*(5), 628–647.

Bers, M., & Portsmore, M. (2005). Teaching partnerships: Early childhood and engineering students teaching math and science through robotics. *Journal of Science Education and Technology, 14*(1), 59–73.

Bers, M. U., Flannery, L., Kazakoff, E. R., & Sullivan, A. (2014). Computational thinking and tinkering: Exploration of an early childhood robotics curriculum. *Computers & Education, 72*, 145–157.

Bers, M. U., Ponte, I., Juelich, C., Viera, A., & Schenker, J. (2002). Teachers as designers: Integrating robotics in early childhood education. *Information Technology in Childhood Education Annual, 1*, 123–145, Association for the Advancement of Computing in Education (AACE).

Bers, M. (2008). Engineers and storytellers: Using robotic manipulatives to develop technological fluency in early childhood. In O. Saracho & B. Spodek (Eds.), *Contemporary perspectives on science and technology in early childhood education* (pp. 105–125). Charlotte, NC: Information Age Publishing.

Black, P., Harrison, C., Lee, C., & Marshall, B. (2008). *Working inside the black box: Assessment for learning in the classroom.* London: GL Assessment.

Brookhart, S. M. (2013). *How to create and use rubrics for formative assessment.* Alexandria, VA: Assn/Supervision & Curric. Dev.

Caitlin, D. (2012). Maximising the effectiveness of educational robotics through the use of assessment for learning methodologies. Paper presented at the 3rd International Workshop Teaching Robotics, Teaching with Robotics: Integrating robotics within school curriculum. Riva del Garda, Italy.

Cannon-Bowers, J., & Salas, E. (2001). Reflections on shared cognition. *Journal of Organizational Behavior, 22*(2), 195–202.

Caviglioli, O., Harris, I., & Tindall, B. (2002). *Thinking skills and Eye Q: Visual tools for raising intelligence.* Stafford: Network Educational Press Ltd.

Cejka, E., Rogers, C., & Portsmore, M. (2006). Kindergarten robotics: Using robotics to motivate math, science, and engineering literacy in elementary school. *International Journal of Engineering Education, 22*(4), 711–722.

Chalmers, C. (2009). *Primary students' group metacognitive processes in a computer supported collaborative learning environment.* Ph.D. thesis, Queensland University of Technology. http://eprints.qut.edu.au/29819/

Chalmers, C. (2013). Learning with FIRST LEGO League. In *Society for Information Technology and Teacher Education (SITE) Conference (25–29 March, 2013)* (pp. 5118–5124). New Orleans, Louisiana: Association for the Advancement of Computing in Education (AACE).

Chalmers, C., & Nason, R. (2005). Group metacognition in a computer-supported collaborative learning environment. In Looi, C., Jonassen, D. H., & Ikeda, M. (Eds.), *Towards Sustainable and Scalable Educational Innovations Informed by the Learning Sciences: Sharing Good Practices of Research, Experimentation and Innovation* (pp. 35–41). Amsterdam: IOS Press.

Chambers, J. M., Carbonaro, M., & Murray, H. (2008). Developing conceptual understanding of mechanical advantage through the use of Lego robotic technology. *Australasian Journal of Educational Technology, 24*(4), 387–401.

Chambers, J., Carbonaro, M., & Rex, M. (2007). Scaffolding knowledge construction through robotic technology: A middle school case study. *Electronic Journal for the Integration of Technology in Education, 6*, 55–70.

Charles, R. (2005). Big ideas and understandings as the foundation for elementary and middle school mathematics. *Journal of Education Leadership, 7*(3), 9–24.

Darling-Hammond, L., Austin, K., Cheung, M., & Martin, D. (2008). Session 9: Thinking about thinking—Metacognition. In *The Learning Classroom: Theory into Practice* (pp. 157–172). Pal Alto, CA: Stanford University School of Education. http://www.learner.org/courses/learningclassroom/

de Bono, E. (1985). *Six thinking hats: An essential approach to business management*. New York: Little, Brown and Co.

Earl, L. (2003). *Assessment as learning: Using classroom assessment to maximise student learning*. Thousand Oaks: CA, Corwin Press.

Eberle, B. (1997). *Scamper worksheet*. USA: Prufrock Press. http://bmgi.org/toolstemplates/scamper-worksheet

Fathulla, K., & Basden, A. (2007). What is a diagram? In *Proceedings of IV07: The 11th International Conference on Information Visualisation (4–6 July 2007)*, Zurich, Switzerland. IEEE Society Press. www.graphicslink.co.uk/IV07/

Fiore, S., & Schooler, J. W. (2004). Process mapping and shared cognition: Teamwork and the development of shared problem models. In E. Salas & S. Fiore (Eds.), *Team cognition: Understanding the factors that drive process and performance* (pp. 133–152). Washington, DC: American Psychological Association.

Fortus, D., Dershimer, R. C., Krajcik, J., Marx, R. W., & Rachel Mamlok-Naaman, R. (2004). Design-based science and student learning. *Journal of Research in Science Teaching, 41*(10), 1081–1110.

Grubbs, M. (2013). Robotics intrigue middle school students and build stem skills. *Technology and Engineering Teacher, 72*(6), 12–16.

Hamilton, E., Lesh, R., Lester, F., & Brilleslyper, M. (2008). Model-eliciting activities (MEAs) as a bridge between engineering education research and mathematics education research. *Advances in Engineering Education, 1*(2), 1–25.

Hamner, E., Lauwers, T., Bernstein, D., Nourbakhsh, I., & DiSalvo, C. (2008). Robot diaries: Broadening participation in the science pipeline through social technical exploration. In *Proceedings of the AAAI Spring Symposium on Using AI to Motivate Greater Participation in Computer Science*, Palo Alto, CA. www.cs.cmu.edu/~illah/PAPERS/aaaiss08rd.pdf

Harlen, W. (Ed.). (2010). *Principles and big ideas of science education*. Hatfield, Herts: Association of Science Teachers.

Hmelo-Silver, C. E. (2004). Problem-based learning: What and how do students learn? *Educational Psychology Review, 16*(3), 235–266.

Jasparro, R. J. (1998). Applying systems thinking to curriculum evaluation. *National Association of Secondary School Principals. NASSP Bulletin, 82*(598), 80–86. http://gateway.library.qut.edu.au/login?, http://search.proquest.com/docview/216028552?accountid=13380du

Johnson, D., & Johnson, R. (Eds.) (2004). *Assessing students in groups: Promoting group responsibility and individual accountability*. Newberry Park, CA: Corwin Press.

Kokotovich, V. (2008). Problem analysis and thinking tools: An empirical study of non-hierarchical mind mapping. *Design Studies, 29*(1), 49–69.

Lane, A. (2013). A review of diagramming in systems practice and how technologies have supported the teaching and learning of diagramming for systems thinking in practice. *Systemic Practice and Action Research, 26*(4), 319–329.

Lesh, R., & Clarke, D. (2000). Formulating operational definitions of desired outcomes of instruction in mathematics and science education. In A. E. Kelly & R. A. Lesh (Eds.), *Handbook of research design in mathematics and science education* (pp. 113–149). Mahwah, NJ: Lawrence Erlbaum.

Lesh, R., Cramer, K., Doerr, H., Post, T., & Zawojewski, J. S. (2003). Model development sequences. In R. Lesh & H. Doerr (Eds.), *Beyond constructivism: Models and modeling perspectives on mathematics problem solving, learning, and teaching* (pp. 35–54). Mahwah, NJ: Lawrence Erlbaum Associates.

Lesh, R., & Doerr, H. M. (Eds.). (2003). *Beyond constructivism*. Mahwah, NJ: Lawrence Erlbaum Associates.

Lesh, R. A., Kelly, A. E., & Yoon, C. (2008). Multi-tiered design experiments in mathematics, science and technology education. In A. E. Kelly, R. A. Lesh, & J. Y. Baek (Eds.), *Handbook of design research methods in education* (pp. 131–148). New York: Routledge, Taylor & Francis.

Norton, S. J., McRobbie, C. J., & Ginns, I. S. (2007). Problem solving in a middle school robotics design classroom. *Research in Science Education, 37*(3), 261–277.

Papert, S. (1980). *Mindstorms. Children, computers and powerful ideas.* New York: Basic Books.

Puntambekar, S., & Kolodner, J. L. (2005). Toward implementing distributed scaffolding: Helping students learn science from design. *Journal of Research in Science Teaching, 42*(2), 185–217.

Rockland, R., Bloom, D. S., Carpinelli, J., Burr-Alexander, L., Hirsch, L. S., & Kimmel, H. (2010). Advancing the "E" in K-12 STEM Education. *Journal of Technology Studies, 36*(1), 53–64.

Rogers, C. B. (2012). Engineering in kindergarten: How schools are changing. *Journal of STEM Education: Innovations and Research, 13*(4), 4–9.

Rogers, G., & Wallace, J. (2000). The wheels on the bus: Children designing in an early years classroom. *Research in Science & Technological Education, 18*(1), 127–136.

Rusk, N., Resnick, M., Berg, R., & Pezalla-Granlund, M. (2008). New pathways into robotics: Strategies for broadening participation. *Journal of Science Education and Technology, 17*(1), 59–69.

Sadler, D. R. (1989). Formative assessment and the design of instructional systems. *Instructional Science, 18*(2), 145–165.

Sadler, P. M., Coyle, H. P., & Schwartz, M. (2000). Engineering competitions in the middle classroom: Key elements in developing effective design challenges. *The Journal of the Learning Sciences, 9*(3), 299–327.

Scardamalia, M. (2002). Collective cognitive responsibility for the advancement of knowledge. In B. Smith (Ed.), *Liberal education in a knowledge society* (pp. 67–98). Chicago, IL: Open Court.

Silk, E. M. (2011). *Resources for learning robots: Environments and framings connecting math in robotics.* Ph.D. thesis, University of Pittsburgh. http://www.education.rec.ri.cmu.edu/content/educators/research/files/SilkEliM2011.pdf

Silk, E. M., Higashi, R., Shoop, R., & Schunn, C. D. (2010). Designing technology activities that teach mathematics. *The Technology Teacher, 69*(4), 21–27.

Stiggins, R. J., Arter, J. A., Chappuis, J., & Chappuis, S. (2007). *Classroom assessment for student learning: Doing it right-using it well.* Upper Saddle River, NJ: Pearson Education Inc.

Sullivan, F. R. (2008) Robotics and science literacy: Thinking skills, science process skills and systems understanding. *Journal of Research in Science Teaching, 45*(3), 373–394.

Sullivan, F. R., & Heffernan, J. (2016). Robotic construction kits as computational manipulatives for learning in the STEM disciplines. *Journal of Research on Technology in Education, 48*(2), 105–128

Welch, M. (1999). Analyzing the tacit strategies of novice designers. *Research in Science & Technological Education, 17*(1), 19–34.

Wendell, K. B., & Kolodner, J. L. (2014). Learning disciplinary ideas and practices through engineering design. In A. Johri & B. M. Olds (Eds.), *Cambridge handbook of engineering research education* (pp. 243–265). Cambridge, UK: Cambridge University Press.

Wendell, B. K., & Rogers, C. (2013). Engineering design-based science, science content performance, and science attitudes in elementary school. *Journal of Engineering Education, 102*(4), 513–540.

Wiggins, G. B., & McTighe, J. (2005). *Understanding by design* (2nd ed.). Heatherton, Vic: Hawker Brownlow Education.

Williams, D. C., Ma, Y., Prejean, L., Ford, M. J., & Lai, G. (2007). Acquisition of physics content knowledge and scientific inquiry skills in a robotics summer camp. *Journal of Research on Technology in Education, 40*(2), 201–216.

Yildirim, T. P., Shuman, L., & Besterfield-Sacre, M. (2010). Model-eliciting activities: Assessing engineering student problem solving and skill integration processes. *International Journal of Engineering Education, 26*(4), 831–845.

Chapter 3
Combatting the War Against Machines: An Innovative Hands-on Approach to Coding

Jacqui Chetty

Abstract The twenty-first century has been an era of technological advances that has surpassed previous decades. This is largely due to the level of innovation in the fields of artificial intelligence, robotics and automation. However, learners are often reluctant to choose computer programming (coding) as a subject due to its perceived difficulty. Nevertheless, it is also well known that learners who are introduced to computer programming at a young age become the computer science university graduates of tomorrow. Learners' hesitancy towards computer programming is due to the complex, abstract nature of the discipline. To this end, innovative tools are proving useful for learners to overcome such barriers. This chapter provides effective strategies to teach the fundamental concepts of computer programming using robotics, specifically Lego Mindstorms robots. The approach taken is hands-on, student-centred and visual. Learners develop coding solutions through designing and coding real-world problems, visually correcting their imprecisions. This chapter includes learning activities and practical examples from case studies. The inequality of women in the workplace, especially women in IT, is also addressed. A discussion around effective approaches to teaching girls' coding is included as research indicates that girls' learning requirements for coding are different to those of boys.

Keywords Lego Mindstorms robotics · Innovative tools · Computer programming · Learner-centred learning · Authentic learning

3.1 Introduction

Learners enrolled for a computer programming module for the first time often find it challenging to understand the fundamental concepts surrounding the discipline. Equally, educators find it difficult to teach such learners. Research indicates that

J. Chetty (✉)
University of Johannesburg, Johannesburg, Gauteng, South Africa
e-mail: jacquic@uj.ac.za

© Springer International Publishing AG 2017   59
M.S. Khine (ed.), *Robotics in STEM Education*,
DOI 10.1007/978-3-319-57786-9_3

traditional pedagogical approaches do not lend themselves towards a positive experience for both learner and educator (Lister 2011). Given that pedagogical approaches are a powerful determinant as to whether a learner will be successful in a programming module, it is worthwhile investigating the variety of pedagogies and related tools available, to determine whether there is an approach that could adequately scaffold learners studying such modules. Research indicates that using games as a pedagogical tool to teach learners, the art of algorithmic problem-solving as well as programming has successfully been implemented to scaffold learners (Badger 2009).

The idea of using games to teaching fundamental computer programming concepts is not new (Lawhead et al. 2002). Using games as a learning tool is advocated as games have the potential to positively contribute to successful learning (Piteira and Haddad 2011). Lego Mindstorms robots is one such game that provides an innovative teaching tool for building learners' computer programming skills. Amongst others, the game provides two necessary elements for learning, namely understanding and motivation (Piteira and Haddad 2011). It provides a platform for learners to build, reinforce and practice fundamental computer programming concepts, while adding an element of fun. Lego Mindstorms scaffolds learners' learning because it uses action instead of explanation; accommodates a variety of learning styles and skills; reinforces mastery skills; provides an opportunity to practice; and affords an interactive, decision-making context.

Lego Mindstorms robots may prove to be an effective teaching tool to scaffold learners as it assists them with refining the art of computational thinking and enables the learning around the design of complex algorithms. The scaffolding becomes an instrument for educators to 'bridge' learners who are navigating a new discipline, as they clasp frantically to fragile knowledge in their minds. Structured scaffolding that is maintained for a reasonable period of time allows the fragile knowledge to solidify and become entrenched into the mind of the learner.

This chapter provides an overview of Lego Mindstorms robots as a tool that can be used to teach or reinforce fundamental computer programming concepts. Guidelines regarding teaching, managing and assessing such lessons are also deliberated. Gender similarities and differences when learning computer programming are also highlighted so that educators are aware of how to ensure that both genders receive appropriate teaching and learning.

3.2 Difficulties Faced by Learners Learning a Computer Programming Language

The skills expected for computer programming are complex, and learners worldwide find it very difficult to solve problems (Mead et al. 2006; Organisation for Economic Co-Operation and Development-OECD 2004). The problem arises as learners need to articulate a problem into a programming solution (Garner et al. 2005; Lahtinen et al. 2005) by combining syntax and semantics into a valid program

(Winslow 1996) through the construction of mechanisms and explanations (Soloway 1986). In order to achieve this, learners need to be able to apply fundamental computer programming concepts (Garner 2005; Robins et al. 2003) and understand abstract concepts (Lahtinen et al. 2005). When learners are faced with trying to absorb and understand too many new concepts at one time, their working memory may become overloaded. Their overloaded working memories make it very difficult for them to understand the concepts taught to them. The idea of working memory and load capacity is also known as cognitive load theory (Mason et al. 2015).

3.3 Cognitive Load

Humans are limited to a working memory capacity that is strictly bounded and relatively small (Mason et al. 2015). This means that due to our limited working memory capacity, our memories can become overloaded and our cognitive performance can decline. This is particularly true when novice learners are faced with the fundamentals associated with computer programming concepts as these concepts are fraught with abstract ideas or higher order thinking skills (HOTS). Such concepts are often layered, one on top of the other, before a learner is able to design and construct computer programs. Given that such learners are new to the discipline (novices), their cognitive load increases exponentially, often exceeding their critical threshold level of cognitive capacity (Mason et al. 2015).

It is therefore not unexpected for research to indicate that the results linked to computer programming modules aimed at novices more often than not have a particularly high failure rate. When such learners are presented with a subject, such as computer programming, they struggle as their cognitive load is pushed to capacity. These learners often cannot adapt as they are expected to learn concepts that require abstract reasoning, also known as computational thinking (Bower and Falkner 2015).

3.4 Computational Thinking

Computational thinking (CT) can be defined as the ability of a learner to develop problem-solving strategies and techniques that assist in the design and use of algorithms and models (Falkner et al. 2015). According to Lister (2011), such thinking needs time to develop. In fact, most learners possess limited CT in the early stages of their lives, but such skills should develop and mature, given that learners are educated and receive formal training (Lister 2011). However, many learners' computational thinking is not developed, and when they are confronted with a discipline such as programming, they are unable to think in a computational manner. Lego Mindstorms robots provides an excellent opportunity to develop CT through the use of problem-solving. Lego Mindstorms and problem-solving are tantamount providing a rich environment that can develop programming skills.

3.5 Problem-solving

Teaching computer programming using a problem-solving approach has become popular over the last decade (Falkner and Palmer 2009; Guillory 2011; Levy and Iturbide 2011; Organisation for Economic Co-Operation and Development-OECD 2004). The idea is that if a student learns how to solve one type of problem using a problem-solving approach, that student should be able to solve other problems of a similar nature, regardless of programming languages (Pears et al. 2009; Winslow 1996). Problem-solving provides an opportunity for students to construct programs using English-like lines of instructions as English is a language that is familiar to them. Expressing algorithmic instruction becomes easier as students are learning on the 'fringes' of what they know. The language is the known and the problem-solving (algorithmically) is the unknown. Lego Mindstorms instructions are English-like, enticing learners to create instructions that are relevant and applicable to the problem at hand.

3.6 Pedagogies Aimed at Computer Programming

Pedagogical approaches to teaching and learning can follow many paths, as there is an abundance of educational paradigms and theories available to teachers and educators alike. Some examples of such paradigms relate to behaviourism, cognitivism and humanism (Knowledgebase 2012). However, social constructivism is a philosophy that is well suited to learning computer programming.

3.7 Social Constructivism and Computer Programming

Social constructivism is a philosophy and a learning theory that has established itself in the last decade (Karagiorgi and Symeou 2005). It is a didactic approach that provides an opportunity for learning to take place in a social, interactive and collaborative manner. Central to this approach is that teaching and learning is an active process of constructing knowledge in a social setting, where students learn collaboratively. The idea is that during the classroom experience, new knowledge is constructed in a social setting. The new knowledge is based on students' past experiences and the hypotheses of the environment in which they learn. An essential aspect concerns how knowledge is being shaped through the use of symbolic tools, such as language (Kozulin and Gindis 2003). The intention is that students develop skills, such as reasoning, problem-solving, the development of higher mental processes and metacognition (Kozulin and Gindis 2003).

Computer programming is a discipline that requires critical thinking, teamwork and the development of software solutions (active process of constructing software solutions). Social constructivism could prove to be an ideal pedagogical approach in this context. Scholars, such as Vygotsky, Piaget and Bruner, believed learning to be an active contextualised process that requires social negotiation (Bruner 1960; Corney et al. 2012; Vygotsky 1978). The process of developing software solutions emulates the philosophies surrounding social constructivism (Chetty 2016).

Parallels can be drawn between social constructivism and computer programming. Computer programming solutions are often developed through a process of discussion. A team of individuals, in collaboration with one another, are required to solve problems and find solutions (Farrell 2010). For example, a computer program is designed and developed using a formal computer programming language, such as $C^{\#}$. The design and the development of the program is a process that requires analogical reasoning, critical thinking, problem-solving and social negotiation. This process often takes place in collaboration with team members, where the design and the development of a program to be constructed requires team members to discuss and negotiate designs, workload, workflow and optimal solutions.

Although social constructivism is an underlying paradigm that can form a foundation for teaching, authentic learning is a platform that can transform the theoretical constructs of social constructivism into practice.

3.8 Authentic Learning Shaping the Design

Bruner stated that there is a real difference between learning about a domain (such as computer programming) and learning to be a computer programmer (Bruner 1960). While facts and knowledge can be taught, these only take on meaning and relevance when students discover the benefits of actively participating during learning as opposed to passively listening in classes (Lombardi 2007). When students actively participate in learning, they learn to 'do', they collaborate with others and they form communities of practice.

Herrington's Authentic Learning Framework, which comprises nine elements (Herrington and Parker 2013), is an excellent example of how authentic learning can be incorporated into the classroom. Although the framework consists of many elements, this chapter considers two elements that can be included when teaching Lego Mindstorms robots (although others could also be included).

3.9 Authentic Context

An authentic context is a physical environment that reflects or mimics the way in which the constructed knowledge is to be used and produced in the real world (Herrington 2006). Some scholars believe that it is the only context in which learning becomes meaningful. These scholars were the pioneers responsible for developing an authentic learning model that bridged the gap between theoretical learning in the classroom and real-life application in the work environment (Herrington 2006). The learning environment can be described as one in which the design of the classroom setting preserves the complexity of the real-life setting. Furthermore, such a setting provides purpose and motivation for learners. It also provides an environment where educators and learners alike can discuss and explain ideas within the context of the real-life situation, making no attempts to fragment or simplify the environment (Herrington 2013a).

Given the notion of what an authentic context is, what would constitute a typical real-world setting for a computer programmer? In order to illustrate the setting, it is important to describe the activities and habits of a computer programmer during any given day so that the setting can reflect this. Computer programmers work on an individual basis as well as within a team. In both of these situations, their goal is to solve problems and develop application software solutions (Career One Stop 2008). Therefore, the workspace environment should reflect this. The working space may consist of a personal space that allows programmers a certain amount of privacy, free from interruption (Barker 2012). In this space, they critically analyse, have time to think creatively and develop programming solutions. The need will arise for them to work collaboratively, and a communal space will then be required. This space should allow them to communicate effectively with one another. Whiteboards and other essentials should be available so that they can discuss and visualise solutions in order to find an optimal solution. The idea is to provide an environment for learners to 'to be programmers' by building Lego Mindstorms solutions like they would be in the real world.

3.10 Authentic Task

Researchers and experts worldwide agree that an authentic learning activity represents a problem that has real-world relevance, is ill-defined and needs to be completed over a period of time (Brannock et al. 2013; Herrington 2006, 2013b; Lombardi 2007). Real-world relevance is concerned with problems that match the real-world tasks of professionals in practice. Such problems are normally 'messy' or ill-defined. Ill-defined problems are problems that when described to students are open to interpretation, as opposed to problems that are developed step-by-step. Instead of being highly prescriptive, ill-defined problems provide an opportunity for learners to identify the steps needed to complete the activity (Herrington et al. 2006). As ill-defined problems are more complex, learners need a longer period of time to complete such activities. The longer time period also allows learners to reflect on the choices that they are making regarding the solution, and this enhances their metacognitive skills (Lombardi 2007).

Authentic learning activities provide an opportunity for learners to construct new knowledge instead of reproducing existing knowledge. In order to achieve this, learners are provided with multiple sources from which they can draw information, examine the problem from many angles, distinguish relevant information from irrelevant information and formulate a product (Lombardi 2007). For example, learners can be asked to develop software solutions, given a real-world problem. These tasks should be completed in collaboration, where learners are given the opportunity to discuss problems, ideas and solutions, thus learning from one another, before completing the task. Authentic tasks can easily be incorporated into a Lego Mindstorms environment providing endless imaginative projects.

3.11 LEGO Mindstorms Robots: A Tool to Scaffold Learning

Lego Mindstorms robots have become a popular pedagogical tool to teach and learn introductory computer programming concepts (Lawhead et al. 2002; Lui et al. 2010). The emphasis is on the word 'tool', where robots create a rich environment that provides a platform for novices as well as experienced educators to implement a laboratory experience for learners to learn programming skills in an interesting, unique and challenging manner. In effect, Stein (1998) challenges the computer science teaching community to move from the premise that computation is calculation to the notion of computation is interaction. Robots would be a natural way to explore such a concept.

Lego Mindstorms robots form part of Lego education and can be bought through a representative responsible for retailing such toys. The Mindstorms consists of building components, a programmable brick, active sensors and motors. There is software for which both graphical user interface (GUI) and command line interfaces are available. The robots, together with their associated interfaces, provide an opportunity for educators to transform classrooms into rich laboratory or software studios, where learners can experience learner-centred learning, collaborative learning and peer-to-peer programming experimentation (Yamazaki et al. 2015). This environment provides an opportunity for learners to 'put their programming skills to the test' as what they program comes to life through the Lego Mindstorms robot. They can visually understand 'what works', 'what does not work' and 'why'. Figure 3.1 shows a robot that is about to perform a task.

Lego Mindstorms robots provides an opportunity for learners to understand fundamental computer programming concepts that are, by their very nature, abstract (deRaadt 2008). These concepts are not analogies with the real world (Piteira and Haddad 2011). However, the Lego Mindstorms programming hides the abstract complexity by providing a fun, click and drag, prompting interface to assist with such analogies.

Introducing Lego Mindstorms robots provides a unique opportunity to transform a classroom environment that can create a degree of motivation in which (Piteira and Haddad 2011):

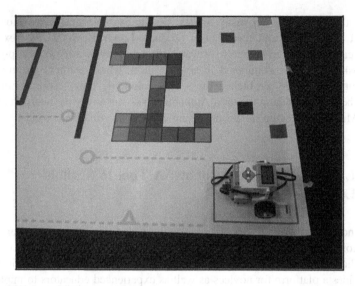

Fig. 3.1 EV3 Lego Mindstorms robot

(a) Learners are given an opportunity to 'grapple' with real-world problems;
(b) The Lego Mindstorms robot becomes a learning tool that can scaffold learners;
(c) Fragile knowledge of abstract programming concepts can be reinforced; and
(d) Learners are given an opportunity to experiment, explore and enjoy programming.

3.12   The Motivation Factor

Research indicates that emotions, such as hope, anger, relief, anxiety and boredom, are significantly related to motivation, learning strategies, cognitive resources, self-regulation, and academic achievement, as well as personality and classroom antecedents (Pekrun et al. 2002). According to Jenkins, motivation in particular is a crucial component related to learners' success. Although motivation is difficult to quantify, Jenkins has identified expectancy and value as two factors, which when multiplied can predict learners' motivation (Jenkins 2001). Expectancy is related to the extent to which learners feel that they are able to succeed. Value is related to what they expect to gain. For example, confident learners who feel that they are able to succeed will attach a value or goal related to high marks. They will most likely score high in the area of motivation as motivation = expectancy * value (Jenkins 2001).

A motivated learner would therefore experience emotions related to hope, enjoyment and pride, whereas an unmotivated learner would experience emotions related to anger, frustration, anxiety and boredom. Lego Mindstorms robots provides an opportunity for learners to experiment and explore. The idea of learning through play is an effective tool to create personal motivation and satisfaction of learning (Piteira and Haddad 2011).

3.13 A Hands-on Approach to Teaching Lego Mindstorms Robotics

Lego Mindstorms robots can be purchased as Lego Mindstorms NXT or the newest version, namely Lego Mindstorms EV3 (lego.com). The EV3 has step-by-step instructions to build a variety of robots, as well as all the components needed to execute a variety of programs so that the robot can perform actions. Figure 3.2 shows some of the components that make up the EV3 robot.

The EV3 brick has a LINUX operating system, memory, ports, USB adaptors and a power supply. There are ports A to D and ports 1 to 4. Each port is an entry/exit point where cables are attached and link to the motors and other sensors. The brick provides the necessary hardware and software for learners to write executable programs. Once learners have built the robot using the Lego pieces, the programmable brick and other components, learners can start writing and executing programs.

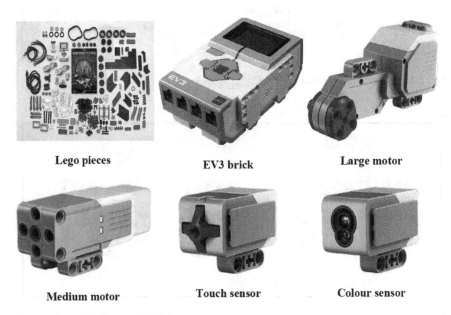

Lego pieces   EV3 brick   Large motor

Medium motor   Touch sensor   Colour sensor

Fig. 3.2 Lego Mindstorms EV3 kit

The programs can be written using the EV3 brick interface or a software application can be downloaded for free (EV3 software). The EV3 software has a powerful interface, as shown in Fig. 3.3, and makes use of a click and drag approach to developing programs. The commands, shown in Fig. 3.3, are grouped by colours, for example green depicts action. All the commands are visible, and learners are not left guessing as to how a program can be built.

The software makes use of a click and drag approach, where learners locate a command and drag it onto the palette. Each command clicks into place, similar to that of puzzle pieces, as shown in Fig. 3.4. Figure 3.4 depicts a robot moving forward at a certain speed over a period of time. The robot then stops once that time period ceases to exist.

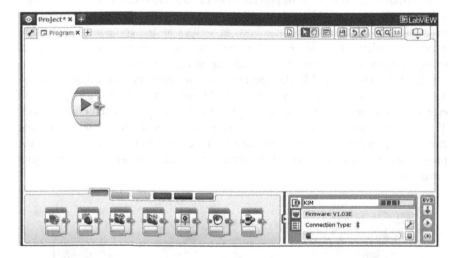

Fig. 3.3 EV3 software interface

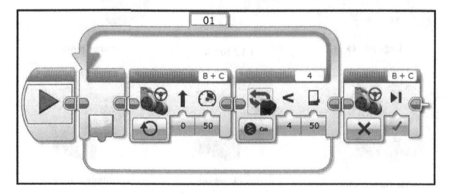

Fig. 3.4 Lego Mindstorms code depicting a loop

The following section provides curriculum guidance to educators that choose to include Lego Mindstorms robotics as part of the classroom learning. The learning discussed next consists of planning, managing and assessing Lego Mindstorms robot's curriculum.

3.14 Planning Lego Mindstorms Lessons

All planning, regarding of subject content, follows a particular pattern (Simmons and Hawkins 2015). For example, lesson plans are developed for a single lesson, collected lesson plans for a number of weeks and long-term plans. This chapter focuses on a single lesson as well as a combination of lessons for a number of weeks as Lego Mindstorms robots is generally taught over a period of a few weeks.

3.14.1 Lesson Plans

A lesson plan is a planning tool that contains all the information as well as the decisions that need to be considered before teaching (Simmons and Hawkins 2015). It consists of learning aims and learning objectives that constitute that lesson. The lesson plan provides the overall context for a lesson as well as activities, outputs and assessment criteria for those outputs. A Lego Mindstorms robots lesson plan should consist of a similar structure.

Lego Mindstorms robots provides a hands-on approach to learning and this means that even the best planned lessons can collapse. This is due to a number of reasons but one reason may be due to the use of technology as the focus of instruction. Many of us have experienced technology failing just as teaching is about to commence, even when many practice rounds have been put in place. For example, a cable can break down or a learner downloads a program but then executes the incorrect program. Teaching programming using innovative tools does increase the risk of lessons failing to achieve the desired outputs, and this can cause much stress to the educator. However, it is important to remember that learners and educators alike learn best when lessons do not go according to plan. A learner that spends a period of time problem-solving why a program does not execute is unlikely to make the same mistake again. An educator that encounters a cable problem more than ten times quickly learns how to solve that particular problem. I have experienced a few common errors that can occur.

Practical Tips

- The EV3 firmware is incompatible with the software application. This is easily rectified by downloading the correct version of firmware;

- Cables used to download programs from the software application to the EV3 brick can malfunction;
- A learner programs a robot to move in a straight line; however, the robot turns in circles. This is normally due to a cable (connected to a sensor on the robot) that is touching a wheel of the robot; and
- The learner has downloaded a program but executes the incorrect program from the EV3 brick. This can be rectified by ensuring that learners name their programs properly.

3.14.2 Planning to Plan

Prior to the development of a lesson plan, it is very important to consider the number of learners in a group as this affects classroom layout, as well as the seating plan. In most instances, it is advisable for learners to work in groups due to, firstly, the expense of the Lego Mindstorms robots. Secondly, Lego Mindstorms robots provides an opportunity for learners to work collaboratively, which is generally appreciated as an excellent approach to learning.

Practical Tip

- A group of two learners is optimal as a group that consists of any more than two learners often means that the other learners in the team do not get to participate —there are not enough tasks for everyone.

3.15 Lego Mindstorms Lesson Plans

3.15.1 Lesson Plan #1

The first lesson plan is relatively straightforward as learners are expected to build the Lego Mindstorms robot. Table 3.1 provides an example of the first lesson plan.

Practical Tips

- The batteries are normally inside the EV3 brick but check;
- Put the EV3 brick on charge while building the robot;
- It may be difficult to establish a timeline for building the robot. It may take from one to two hours; and
- Make sure that the cables are attached to the correct ports/components (i.e. sensors and motors).

Table 3.1 Lesson plan #1

Time	Learner activity	Teacher activity	Learning outcome
1 to 2 h	Provide an opportunity for learners to build the robot in groups where they can discuss and collaborate on their design The building of Lego Mindstorms robots offers learners the opportunity to create their unique bot—ready to be programmed and controlled in the manner that suits them Learners that feel more comfortable with step-by-step, Lego-type instructions can build their bot by choosing a design where the instructions for the building are already given to them Provide them an opportunity to deviate from their model to include a few components of their own	Welcome the class and provide an introductory talk about building the robot Point out the different components that are needed to build the robot Offer learners the opportunity to deviate from the set models by designing and building their own unique bot	The completion of the robot to a satisfactory level

3.15.2 Lesson Plan #2

The second lesson plan involves an explanation of the software application interface, also known as the Programming Canvas. There are a variety of instructions or commands, each grouped and located in a Programming Palette (divided into categories by colour).

The palettes are listed as follows:

Action blocks		Flow blocks		Sensor blocks		Data blocks	
☐	Medium motor	☐	Start	☐	Brick buttons	☐	Variable
		☐	Wait	☐	Colour sensor	☐	Constant
☐	Large motor	☐	Loop	☐	Infrared sensor	☐	Array operations
☐	Move	☐	Switch	☐	Motor rotation	☐	Logic operations
	steering	☐	Loop interrupt	☐	Timer	☐	Math
☐	Move tank			☐	Touch sensor	☐	Round
☐	Display					☐	Compare
☐	Sound					☐	Range
☐	Brick status					☐	Text
	light					☐	Random

Table 3.2 Lesson plan #2

Time	Learner activity	Teacher activity	Learning outcome
1–2 h	The learners will familiarise themselves with the various palettes that form part of the interface Learners must remember the different palettes, such as Action and Flow. Ask learners to make a chart so that they can categorise the palettes, together with their commands. The chart can be created in such a way that it can be used in a later lesson (colour sensor). The robot can detect a colour and sound out the palette, such as Action and Flow (see lesson 4#)	Point out the different palettes that will be required by learners to develop a set of instructions Assist learners in understanding the different categories of blocks and why they are grouped accordingly	Learners are expected to be able to differentiate between the categories of Programming Palettes Learners should be able to name a few blocks from the Action blocks category
15 min	Learners create their first program by clicking and dragging the Move Tank command to the right of the Start command Learners must make sure that the two commands fit together like puzzle pieces	Perform the activity with the learners and remind learners where to find the Move Tank command Show learners how to correctly click the commands in place to form a program Explain to learners that a program is a set of step-by-step instructions performed in sequence	Learners are expected to have put the two commands together

There are two more blocks, namely the Advanced blocks and the My blocks, each consisting of a number of commands. Table 3.2 provides an example of the lesson plan.

Practical Tips

- Get learners to ask each other which palette holds which blocks;
- Point out to learners the difference between the medium and large motors;
- Make sure that learners understand the difference between move steering and move tank;
- Let learners measure the length when the tank moves forward for x—seconds, degrees and rotations; and
- Point out to learners that the Flow blocks/Wait command can be used to create instructions for sensors instead of the Sensor blocks.

3.15.3 Lesson Plan #3

Lesson plan #3 is a continuation of lesson plan #2 and therefore can be completed as part of the same session. This lesson plan consists of learning about the hardware associated with the robot and how learners can examine the hardware using the Programming Canvas. The hardware is directly linked to the programming commands, such as moving the wheels or instructing the robot to pick up colour. It is therefore important that learners understand how the hardware, cables and ports operate as well as connect with one another.

Figure 3.5 displays information about the EV3 brick. EV3 is the name of the brick, although this can be altered. Learners enjoy developing unique names for their EV3 brick. The brick information tab displays information about the brick, such as the battery level, firmware version and how much memory has been used. The Port view tab shows which sensors and motors are attached to the brick, indicated in Fig. 3.6.

The Available Bricks tab connects the brick to the EV3 software. The download button allows transfer of the program from the computer to the EV3 brick.

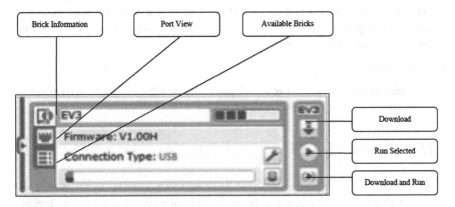

Fig. 3.5 Hardware page tabs on the left and the download/run buttons on the right

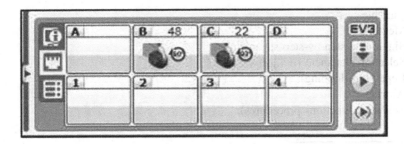

Fig. 3.6 Port view

Table 3.3 Lesson plan #3

Time	Learner activity	Teacher activity	Learning outcome
30 min	Learners observe the various ports, plugging in the different sensors Provide opportunities for learners to collaborate and discuss with one another what the different ports do Allow them to trace the cable leaving an exit point and entering another port Learners download and run a program, observing the activity at the port view	Point out each port associated with a sensor and ensure that learners understand the link between the robot and the port view Provide a small problem for learners to solve to provide an opportunity for learners to download and run a program	Learners are expected to feel comfortable with the various ports Learners should be able to identify which port is connected to which sensor Learners are expected to be able to download and run a program developed in the software application window

This option only allows for a download and does not run the program. The Download and Run option downloads the program and runs the program immediately. The Run Selected will download and run only the blocks that have been selected. This option is useful when fixing problems in a program (Griffin 2014). Table 3.3 provides a lesson plan.

Practical Tips

- Get learners to ask each other which palette holds which blocks—refer them to their charts that they have developed.

3.15.4 Lesson Plan #4

From lesson #4 onwards, the focus of the classroom learning is to provide an understanding of the fundamental concepts associated with computer programming, using Lego Mindstorms robots. These fundamentals[1] are as follows (the completed ones depends on the curriculum):

- Variables;
- Memory;
- Single-line step-by-step statements;
- Selection statement (if … else …);
- Repetition (looping);
- Arrays;
- Methods (functions procedures);

---

[1]An explanation around the fundamental concepts of programming for non-programmers

Table 3.4 Lesson plan #4

Time	Learner activity	Teacher activity	Learning outcome
Time varies for each programming concept taught. On average, each concept should take about 90 min to teach	Learners must discuss, in a group, the programming concept being taught, by providing examples of how that concept is useful or can be used as part of a software solution	Provide an understanding to learners around the programming concept taught, for example variables Provide much opportunity for learners to practice the programming concepts being taught Introduce real-world problems where learners can work in pairs to solve the problem	Learners are expected to understand each programming concept learnt by developing practical solutions that include the programming concept

- Objects; and
- Classes.

Each lesson plan following on from lesson #4 imprints a similar pattern or structure, the only difference being is the computer programming concept taught. Therefore, the layout for lesson #4 can be seen as a template, Table 3.4, that can be used for further lessons (i.e. lesson #5, lesson #6 and so on).

Practical Tips

- Spend a fair amount of time on each programming concept as learners often 'think' they understand until they have to solve a problem that they have never been exposed to before;
- Each of the programming concepts makes use of a variety of palettes. Other than the common Action blocks, the Flow blocks are prominent combined with the Sensor blocks; and
- The Colour sensor, Touch sensor and Infrared sensor are particularly useful for learning about selection (If…/else…) as well as repetition (For… loop/While… loop).

3.16 Managing Lessons

Classroom management can be one of the biggest sources of uncertainty and anxiety for educators and learners alike (Simmons and Hawkins 2015). Even the best planned lesson objectives can be diluted when the act of learning as well as the behaviour of a group of learners is unpredictable. The focus of managing lessons aimed at Lego Mindstorms robots probably falls into two main categories, namely managing the classroom environment and managing the planned lesson.

3.16.1 Managing the Classroom Environment

Experience suggests that it often becomes tricky to manage the behaviour of learners in the classroom as learners find the experience of building, as well as the development of solutions extremely exciting. The classroom experience is a very physical one, where learners move around, solve problems in pairs and are experiencing the joy of learning by doing. The robot moves around, makes sounds, bumps into objects and other robots, and this provides much entertainment amongst learners. It becomes rather difficult to bring calm to the classroom environment, especially when you need learners to take their seats so that a concept can be discussed or reflected upon. The noise levels within the classroom can be very high.

In order for the classroom environment to remain a positive one, it is important that the educator discusses with learners how learning will take place. Some points of discussion can be:

- **Focus learners' attention**—for example, educators can come to an agreement with learners that when a funny word, such as 'beetle juice', is uttered by the educator the learners respond with 'glug, glug, glug' and they all take their seats.
- **Take turns**—for example, educators can, when a pair is grouped, also be known as pair programming (Preston 2005), by allocating one learner as being the driver (typing in the code) and the other learner as being the navigator (observe/correct the driver). With each activity, the driver and the navigator are swopped around. Proper roles may reduce arguing about whose turn it is.
- **Listen and participate in class discussion**—for example, educators can ask a pair to demonstrate their programming solution and discuss the manner in which the solution was developed. Other pairs can rate the solution being demonstrated.
- **Length of time allocated to lessons**—educators must be aware that although learners are very engaged and the time allocated to participate within a lesson seems to go quickly, learners do tire and small breaks are necessary. It is easy to forget about providing a break when learners and the educator are very absorbed within an activity.

3.16.2 Managing the Planned Lesson

The adage 'failing to plan is planning to fail' cannot be more true, especially within a teaching and learning environment. It is very important, and maybe more so when making use of physical objects within a lesson plan, to be vigilant about planning. Carefully crafted learning objectives, writing lesson plans, developing problems for learners to solve and presentations regarding some part of the curriculum provide a structure to educators and communicate confidence and experience to learners.

- **Learning objectives**—other than developing objectives regarding the programming environment for Lego Mindstorms robots, the learning objectives should reflect fundamental programming concepts, linked to the educators' curriculum. There are a variety of books, such as The Art of Lego Mindstorms EV3 Programming and The Lego Mindstorms EV3 Discovery book, to name a few, that are very helpful.
- **Lesson plans**—these can follow the structure as seen above in Tables 3.2, 3.3 and 3.4.
- **Problems for learners to complete**—there is such a wide variety of problems that can be presented to learners. Regardless of the type of problem, real-world problems can provide an opportunity for learners to develop solutions within their world context. For example, ask learners to develop a solution where the robot behaves like an alarm. When an 'intruder' moves within 100 m of the robot, an alarm bell is activated.

3.17 Assessment

The use of assessment for an ICT subject has traditionally been poor although it is improving (Simmons and Hawkins 2015). Unless the ICT subject was receiving specialist ICT teaching, there is often little proof of assessment. Projects are normally completed by learners, and educators may allocate marks in an arbitrary manner, along with feedback, either verbal or written.

However, all assessment should provide a measure of performance against a target standard. Bloom sought to move away from assessing in such a way that learners were compared with one another, and instead compared to a set of objective criteria. Within the computing discipline, such assessment is more often than not a practical one. This can very much be aligned to Bloom's way of assessing as no two programming solutions may be the same. Learners think differently about a problem and often develop different ways of solving a problem.

Assessment should, of course, be aligned to learning outcomes. However, the type of assessment can be formative or summative assessment. For computing solutions, both types of assessment are useful. Whereby formative assessment provides an opportunity for both educator and learner to discuss what needs to be done to solve a problem as well as how to achieve a desired outcome, summative assessment relies on tests and examinations. For example, learners discussing their programming solutions with other students provide a wonderful learning opportunity as solutions always vary. Learners have opportunities to teach and learn from one another.

Formative assessment can be accomplished by the programming pair being asked to develop a solution. Once the solution is developed, the pair demonstrates the end result. The educator can then ask questions from each individual within the pair to verify that each learner did participate in the learning and that learning did take place for both learners.

Summative assessment can be difficult to realise as no two programming solutions are the same. The idea of a rubric that provides generic criteria, from which marks can be allocated, may be considered as useful. However, within this structure, the educator must provide leeway for out-of-the-box lateral thinking and unique solutions that some learners are bound to produce.

3.18 Reflection upon Lessons

While the manner in which learning takes place is important, it is equally important for learners to plan, manage and reflect on their learning (Laskey and Hetzel 2010), also known as metacognition. The term metacognition was originally associated with scholars such as Flavell, Zabrucky and Brown (Bransford et al. 2000). Metacognition encompasses learners consciously and actively understanding their cognitive aptitude and the ability to apply strategies to control cognitive thought processes. Such internal thought patterns extend into learners' daily lives where cognitive tasks are planned, regulated, coordinated and monitored. Learners with good metacognitive skills therefore have the potential to perform better in an academic environment (Schraw and Dennison 1994).

Within a Lego Mindstorms robot classroom, environment metacognition can be accomplished in many ways. For example, learners can be asked to discuss what they learnt within their pairs, within a group or with an educator. One way of doing this is to include the reflective thinking as part of a fun activity. The use of a Koosh ball shown in Fig. 3.7 can be an effective object to encourage reflective thinking. The educator throws or passes the Koosh ball to a learner, and the learner describes something about their classroom learning experience. The Koosh ball is then passed along to another learner and so on. Each learner is provided with an opportunity to reflect, and nobody within the group is to comment or criticise. From experience, learners find this type of activity enjoyable.

Fig. 3.7 Koosh ball often used for therapy

3.19 Teaching Lego Mindstorms to Both Genders

The teaching and learning of programming has always been a subject of much attention, due to the difficulties that learners encounter when learning this discipline. Over the decades, much research has been conducted so that researchers can better understand the ways in which learners learn to program. Exploration regarding gender and whether this influences learning is also an area that has been researched (Burnett et al. 2010; Carter and Jenkins 1999, 2001). Although there are many factors that influence learning to program, research indicates that gender is a significant factor in determining the way in which students approach learning to program (Funke et al. 2015). This section provides insight into teaching different genders and how to provide a teaching and learning environment that best suits each gender.

3.19.1 Gender Differences in Computer Science

In almost all western countries, women are severely underrepresented in the discipline of computer science. Only 20% of an intake within any given department is female (Funke et al. 2015). Given these statistics what can be learnt so that educators are aware of the situation and provide an environment that encourages female learners to enrol for programming courses.

Female learners tend to be less confident and they underestimate their ability (Carter and Jenkins 1999; Funke et al. 2015). Consequently, female learners often have weaker marks, only do what is required of them and they are less fascinated, adopting a pragmatic approach to programming. However, they are also more enthusiastic to seek assistance and readily attend extra tutorials, preferring smaller groups over larger ones. Interestingly, female learners are more consistent, and when confronted with a problem admit that there is a problem before the problem becomes a larger one. Another aspect that affects the learning process is emotions (Chetty and van der Westhuizen 2013). Research indicates that happiness has a positive effect on learning, and anxiety negatively influences the learning process; and the motivation of female learners (Funke et al. 2015). Communication is very important to female learners. Denner et al. show that girls benefit from collaboration where they work together as a pair, one being the driver and the other the navigator.

Male learners are more confident as they depict the typical role model of the male computer scientist. They seem to have more hands-on experience, try and test things out (scientific curiosity), and have more interest probably due to the gaming industry. The result is that male learners often produce better marks. However, male learners are often less structured and often do not admit when there is a problem until the problem at hand is almost insurmountable (Carter and Jenkins 1999). Additionally, they do not readily attend extra tutorials.

3.20 Teaching Lego Mindstorms Robots to Both Genders

Both genders can benefit from learning programming using Lego Mindstorms robots. Table 3.5 provides some practical ideas of how Lego Mindstorms robots can be used to encourage both genders to enjoy the process of learning to program.

Female programmers dominated the industry in the 1960s up to the 1980s, when a decline of female programmers began. This is unfortunate as female programmers are greatly required in this industry due to their unique abilities that yield excellent

Table 3.5 Learning for both genders using Lego Mindstorms robots

Both genders learning programming using Lego Mindstorms robots		
Lego Mindstorms robots learning	Female learners	Male learners
Scientific curiosity	Needs encouragement and can be through the use of real-world problems that are meaningful to them. Provide examples of female programmers, such as the first programmer was a female	Do not need much encouragement as they are naturally curious about robots
Solving smaller problems	Easily solve smaller problems so this may be a way of retaining interest in robots	Benefit from solving smaller problems due to their inability of admitting when a program has a 'bug' until the problem is very large. Encourage male learners to solve small problems in this manner
Student-centred learning / Collaborative learning	Communication is naturally good so encourage female learners to discuss programming problems and solutions as part of a group discussion. This may build up an excitement and happiness around programming, boosting confidence	Male learners may need to be encouraged to discuss learning. Provide an environment where they can discuss in pairs or small groups problems and solutions related to their learning. Be goal-oriented and specific
Small group learning	Any group learning is suitable for female learners. Provide them with an opportunity to discuss their feelings about solving a problem or working on a solution	Male learners may need encouragement and structure, such as providing a driver and a navigator when solving a problem
Real-world problems	Provide practical problems as they relate to their world, each group of students having a different world perspective	Provide practical problems as they relate to their world, each group of students having a different world perspective
Reflection	As female learners communicate and share well, reflection may be an opportunity to enhance their confidence. For example, allow them to share the successes they have experienced	Male learners may require encouragement. Provide a structure whereby you ask them to reflect on an aspect and prevent open-ended questions

programmers. Skills such as attention to detail, meticulousness and an ability to determine and correct 'bugs' within programs quickly, are just a few that come to mind.

3.21 Conclusion

Learners face many difficulties and challenges when presented with programming concepts. However, many of these challenges can be addressed when educators investigate and understand the ways in which learners learn programming best. Although not a silver bullet, good teaching enables learners and provides an environment that encourages learning.

Lego Mindstorms robots may be an effective tool in which the fundamental concepts related to programming can be presented to learners. Mindstorms provides an opportunity to encourage innovative learning styles, such as student-centred collaborative learning. These styles of learning are often successful with learners. Furthermore, Lego Mindstorms robots provides the much needed scaffolding required when teaching programming by presenting difficult programming concepts in a visual, step-by-step way that learners may find easier to grasp. The fun interactive manner in which learning occurs means that students learn from one another.

This chapter provides an overview of how teaching and learning can be accomplished through the use of Lego Mindstorms robots. These robots provide a wonderful opportunity for educators to include much needed scaffolding for an otherwise very difficult discipline such as programming.

References

Badger, M. (2009). *Scratch 1.4 Learn to program while creating interactive stories, games, and multimedia projects using Scratch Beginner's Guide*. Birmingham, Mumbai: PACKT.

Barker, E. (2012). *What do the best programmers have in common?* Retrieved from http://www.bakadesuyo.com/2012/08/what-do-the-best-computer-programmers-have-in/

Bower, M., & Falkner, K. (2015). *Computational thinking, the notional machine, pre-service teachers, and research opportunities*. Paper presented at the Australasian Computer Education (ACE) Conference, Sydney, Australia.

Brannock, E., Lutz, R., & Napier, N. (2013). *Integrating authentic learning into a software development course: An experience report*. Paper presented at the SIGITE'13, Orlando, Florida, USA.

Bransford, J. D., Brown, A. L., & Cocking, R. R. (2000). *How people learn: Brain, mind, experience, and school: Expanded Edition*. Committee on Learning Research and Educational Practice (Ed.) Retrieved from http://www.nap.edu/catalog/9853.html

Bruner, J. (1960). *The process of education*. Cambridge: Harvard University Press.

Burnett, M., Fleming, S., Iqbal, S., Venolia, G., Rajaram, V., Farooq, U., et al. (2010). *Gender differences and programming environments: Across programming populations*. Paper presented at the ESEM '10, Italy.

Career One Stop. (2008). *A day in the life—Computer programmer*. Retrieved from http://www.youtube.com/watch?v=RQ_HdHSpDEg

Carter, J., & Jenkins, T. (1999). *Gender and programming: What's going on?* Paper presented at the ITiCSE'99, Poland.

Carter, J., & Jenkins, T. (2001, September). *Gender differences in programming?* Paper presented at the ITcSCE, Canterbury, UK.

Chetty, J. (2016). *An emerging pedagogy for teaching computer programming: Attending to the learning needs of under-prepared students in university-level courses* (Ph.D. Education Information Systems), University of Johannesburg, Johannesburg, South Africa.

Chetty, J., & van der Westhuizen, D. (2013). *"I hate programming" and other oscillating emotions experienced by novice students learning computer programming*. Paper presented at the EdMedia'13, Canada.

Corney, M., Teague, D., Ahadi, A., & Lister, R. (2012). *Some empirical results for Neo-Piagetian reasoning in novice programmers and the relationship to code explanation questions*. Paper presented at the Australasian Computing Education Conference, Melbourne, Australia.

deRaadt, M. (2008). *Teaching programming strategies explicitly to novice programmers*. (Doctor of Philosophy), University of Southern Queensland.

Falkner, K., & Palmer, E. (2009). *Developing authentic problem solving skills in introductory computing classes*. Paper presented at the SIGCSE'09, Tennessee, USA.

Falkner, K., Vivian, R., & Falkner, N. J. G. (2015). *Teaching computational thinking in K-6: The CSER digital technologies MOOC*. Paper presented at the Australasian Computer Education (ACE) Conference, Sydney, Australia.

Farrell, J. (2010). *Java^{TM} programming* (5 edn). Course Technology, Cengage Learning.

Funke, A., Berges, M., Muhling, A., Hubwieser, P. (2015). *Gender differences in programming: research results and teachers' perception*. Paper presented at the Koli Calling '15, Finland.

Garner, S., Haden, P., Robins, A. (2005). *My program is correct but it doesn't run: A preliminary investigation of novice programmers' problems*. Paper presented at the Australasian Computing Education Conference, Newcastle, Australia.

Griffin, T. (2014). *The Art of Lego Mindstorms EV3 Programming*.

Guillory, B. A. (2011). *Teaching through problem solving*. Retrieved from http://www.slideshare.net/bbieniemy/teaching-through-problem-solving1

Herrington, J. (2006). *Authentic e-learning in higher education: Design principles for authentic learning environments and tasks*. Paper presented at the AACE.

Herrington, J. (2013a). *Authentic contexts set the scene*. Retrieved from http://authenticlearning.info/AuthenticLearning/Authentic_Context.html

Herrington, J. (2013b). *Its the task that matters most*. Retrieved from http://authenticlearning.info/AuthenticLearning/Authentic_Task.html

Herrington, J., & Parker, J. (2013). Emerging technologies as cognitive tools for authentic learning. *British Journal of Educational Technology, 44*(4), 607–615.

Herrington, J., Reeves, T. C., & Oliver, R. (2006). Authentic tasks online: A synergy among learner, task and technology. *Distance Education, 27*(2), 15. doi:10.1080/01587910600789639

Jenkins. (2001). *Teaching programming—A journey from teacher to motivator*. Paper presented at the 2nd Annual LTSN-ICS Conference, London.

Karagiorgi, Y., & Symeou, L. (2005). Translating constructivism into instructional design: Potential and limitations. *Educational Technology & Society, 8*(1), 11.

Knowledgebase, L. T. (2012). Learning-Theories.com. Retrieved from http://www.learning-theories.com/

Kozulin, A., Gindis, B., Ageyev, V. S., & Miller, S. M. (Ed.) (2003). *Vygotsky's educational theory in cultural context*. Cambridge.

Lahtinen, E., Ala-Mutka, K., & Jarvinen, H. (2005). *A study of the difficulties of novice programmers*. Paper presented at the ITiCSE '05, Monte de Caparica, Portugal.

Laskey, M. L., & Hetzel, C. J. (2010). *Self-regulated Learning, metacognition and soft skills: The 21st century learner*. Retrieved from http://www.eric.ed.gov/PDFS/ED511589.pdf

Lawhead, P. B., Bland, C. G., Barnes, D. J., Duncan, M. E., Goldweber, M., Hollingsworth, R. G., et al. (2002). *A road map for teaching introductory programming using LEGO Mindstorms Robots*. Paper presented at the ITiCSE-WSR.

Levy, R. B., & Iturbide, J. A. V. (2011). *A problem solving teaching guide based on a procedure intertwined with a teaching model*. Paper presented at the ITiCSE'11, Darmstadt, Germany.

Lister, R. (2011). *Concrete and other Neo-Piagetian forms of reasoning in the novice programmer*. Paper presented at the Australasian Computer Education Conference, Perth, Australia.

Lombardi, M. M. (2007). Authentic learning for the 21st century: An overview. Retrieved from http://net.educause.edu/ir/library/pdf/ELI3009.pdf

Lui, A. K., Ng, S.C., Cheung, H.Y., & Gurung, P. (2010). Facilitating independent learning with Lego Mindstorms Robots. *acmInroads, 1*(4), 5.

Mason, R., Cooper, G., Simon, & Wilks, B. (2015). *Using cognitive load theory to select an environment for teaching mobile apps development*. Paper presented at the Australasian Computer Education (ACE), Sydney, Australia.

Mead, J., Gray, S., Hamer, J., James, R., Sorva, J., St. Clair, C., et al. (2006). *A cognitive approach to identifying measurable milestones for programming skill acquisition*. Paper presented at the ITiCSE'06, Bologna, Italy.

Organisation for Economic Co-Operation and Development-OECD. (2004). Problem solving for tomorrow's world—First measures of cross-curricular competencies from PISA 2003. Retrieved from http://www.oecd.org/dataoecd/25/12/34009000.pdf

Pears, A., Seidman, S., Malmi, L., Mannila, L., Adams, E., Bennedsen, J., … Paterson, J. (2009). *A survey of literature on the teaching of introductory programming*. Paper presented at the ITiCSE-WGR'07.

Pekrun, R., Goetz, T., Titz, W., & Perry, R. P. (2002). Academic emotions in students' self-regulated learning and achievement: A program of qualitative and quantitative research. *Educational Psychologist, 37*(2). doi:10.1207/S15326985EP3702_4

Piteira, M., & Haddad, S. R. (2011). *Innovate in your program computer class: An approach based on a serious game*. Paper presented at the OSDOC'11, Lisbon, Portugal.

Preston, D. (2005). *Pair programming as a model of collaborative learning: A review of the research*. Paper presented at the CCSC: Central Plains Conference.

Robins, A., Rountree, J., & Rountree, N. (2003). Learning and teaching programming: A review and discussion. *Computer Science Educational Journal, 13*, 137–172.

Schraw, G., & Dennison, R. S. (1994). Assessing metacognitive awareness. *Contemporary Educational Psychology, 19*, 16.

Simmons, C., & Hawkins, C. (2015). *Teaching computing*. London: Sage.

Soloway, E. (1986). Learning to program = learning to construct mechanisms and explanations. *Communications of the ACM, 29*(9), 9.

Stein, L. (1998). What we've swept under the rug: Radically rethinking CS1. *Computer Science Education, 8*(2), 118–129.

Vygotsky, L. (1978). *Mind and society*. Cambridge, MA: Harvard University Press.

Winslow, L. E. (1996). Programming pedagogy—A psychological overview. *SIGCSE Bulletin, 28*(3), 6.

Yamazaki, S., Sakamoto, K., Honda, K., Washizaki, H., & Fukazawa, Y. (2015). *Comparative study on programmable robots as programming educational tools*. Paper presented at the Australasian Computer Education (ACE) Conference, Sydney, Australia.

Chapter 4
The Open Academic Robot Kit

Raymond K. Sheh, Amy Eguchi, Haldun Komsuoglu
and Adam Jacoff

Abstract The Open Academic Robot Kit (OARKit) lowers the barrier of entry into robotics research. A community-driven initiative, it was developed in the context of the RoboCupRescue Robot League competition to advance the state of research in response robotics. All mechanical parts are 3D printable, available off the shelf and/ or, ideally, drawn from a set of common parts. All designs, instructions and source code are available online in easily editable form under an open-source licence. These principles allow the OARKit robots to become powerful tools to encourage collaboration across regions, generations and areas of expertise. The principles that govern this initiative can be applied broadly to other robotics applications that require interdisciplinary skills in order to build complete, useful, interesting research implementations.

Keywords Robotics · Operating system · STEM · Interface · 3D printing

4.1 Introduction

The Open Academic Robot Kit (OARKit) aims to bring interesting research-level robots into the undergraduate and high school classroom. Those wishing to enter into the field of robotics research often face three challenges. These are especially

R.K. Sheh (✉)
Curtin University, Perth, Australia
e-mail: raymond.sheh@curtin.edu.au

A. Eguchi
Bloomfield College, Bloomfield, USA

H. Komsuoglu
Robolit LLC, Philadelphia, USA

A. Jacoff
National Institute of Standards and Technology, Gaithersburg, USA

© Springer International Publishing AG 2017
M.S. Khine (ed.), *Robotics in STEM Education*,
DOI 10.1007/978-3-319-57786-9_4

true for researchers in smaller or non-specialist university departments, under-graduates, secondary schools, hobbyists and smaller companies. The challenges are as follows:

- Building their first complete, working robot system.
- Obtaining relevant, first-hand information about the open problems and existing solutions in the field.
- Finding a community in which to demonstrate and compare their developments.

Over the past decade, we have been developing a variety of initiatives that aim to lower the barrier of entering into response robotics research in particular, and the field of robotics in general. These initiatives include the RoboCupRescue Robot League (RRL) (Sheh et al. 2012) competition and the Response Robotics Summer Schools (and related education events).

Response robots, as used by bomb squads, hazardous material-handling per-sonnel, search and rescue teams and the military, allow responders to address hazardous situations while staying at a safe distance. This domain poses challenges in mechanical design, electronics, communications, computer science, automation, artificial intelligence, human–system interaction, reliability and systems architecture that are common to many other fields of robotics. Examples of response robots appear in Fig. 4.1.

In this chapter, we will discuss the OARKit, which seeks solutions to the first challenge, and two interrelated initiatives that extend our past work in this field and together seek to address these three challenges.

4.2 Closing the Loop on a Working Robot

There are many successful groups around the world performing research into advanced robotics. Many more groups that wish to undertake research in robotics will lack expertise in one or more crucial areas of robotics, such as mechanical engineering, electronic engineering or computer science. Without the requisite expertise at their disposal, designing and building a custom robot of a sufficient level of complexity to perform robotics research can be very challenging and time-consuming.

For groups that lack one or more of these areas of expertise, closing the loop on their first complete and working robot system with which to undertake interesting research often required making a compromise. For example, they could build a simpler robot from scratch, adapt an off-the-shelf kit or using a construction kit. Making solutions a sufficiently good fit to the requirements under these circum-stances can, in some applications, be very challenging or impossible. For example, construction kit robots are often limited in the possible form of structures that can be created, especially when compactness is important. They can also be difficult to share and to leverage the work of others.

Fig. 4.1 Examples of response robots

More recently, high-quality, low-cost rapid prototyping equipment, such as 3D printers and laser cutters, have become readily available. Parts such as microcontrollers, single-board computers, cameras and other components for the maker, hobbyist and radio-controlled model communities, which have traditionally been hard to obtain, are now easily available. We discuss how we leverage these developments in the OARKit (www.oarkit.org), with the aim of making it easier to close the loop on robot systems that can form the basis of interesting robotics research. The two initial robot designs of the OARKit appear in Fig. 4.2.

Fig. 4.2 Emu Mini 2 and the Excessively Complex Six-Wheeled Robot, the first two robot designs in the Open Academic Robot Kit

4.3 Competitions as a Venue for Demonstrating and Comparing Developments

Competitions are a powerful mechanism by which researchers, and especially student researchers, can evaluate the performance of their work. Properly organised competitions also educate researchers on topics such as systems integration, reliability, standardised tests of robot capabilities and statistical significance. For over a decade, we have been conducting competitions such as the RRL that explicitly aim to further research into improving the capabilities of response robots with the aforementioned topics in mind. Figure 4.3 shows the arena in which the RRL is held.

In recent times, other groups have also seen the value of competitions and placed their own spin on the topic. Examples include the Defence Advanced Research Projects Agency (DARPA) Robotics Challenge (DRC), the For Inspiration and Recognition of Science and Technology (FIRST) Robotics Challenge (FRC) and the euRathlon (and preceding European Land Robotics Trials). In conjunction with the recent advances in rapid prototyping and components outlined above, research competitions in this field have become even more accessible. The unique characteristics of the RRL as a driver for research, and the development of a new competition, the RoboCupRescue Rapidly Manufactured Robot Competition will be introduced in this chapter.

Fig. 4.3 The RRL arena. This is a test course for response robots of various shapes and sizes and is built based on a 1.2-m grid. An early version of the Rapidly Manufactured Robot Competition arena is visible embedded on the left

4.4   Disseminating the Challenges and Solutions in the Field

With the advent of the Internet and the many and varied ways of disseminating information, it might seem strange to assert that finding out about the challenges and solutions in this field would be difficult. However, being able to physically interact with hardware and discuss challenges and implementations first-hand provides advantages in finding and developing the right questions to answer. For over a decade, we have run teaching camps and summer schools that bring both the research and end-user communities together. These events are designed to complement other channels such as online resources, conferences and competitions. In this paper, we will also discuss our recent work in bringing together not just researchers, but also responders, industry representatives and members of the maker community to further catalyse work in this field.

4.5   Background

There are many different initiatives that seek to use competitions in order to advance research and development into robotics. Robot competitions, especially in the response robotics arena, are becoming very popular. High-profile examples range from the RoboCupJunior and FIRST competitions at the primary and secondary school levels, the RoboCupRescue Robot League and the Japan Rescue Robotics Contest, to the recent and ongoing DARPA Robotics Challenge and the Robot Summit 2020.

An important distinction that we make in these competitions is in the types of challenges that are set by each competition. Educational competitions, such as RoboCupJunior and FIRST, aim to encourage participants, typically younger students in primary and secondary school, to pursue STEM subjects (Science, Technology, Engineering and Mathematics). While they may contain a research component, they tackle abstract or simulated tasks and generate solutions which do not necessarily solve open, real-world problems directly.

In contrast, competitions at the more senior levels, such as the RoboCupRescue Robot League and the DARPA Robotics Challenge, tend to address challenges that more closely resemble real-world problems for which there are no known or well-implemented solutions. The strength of such an approach is that teams who do well in the competition have also developed technologies that advance the state-of-the-art in capabilities that can be readily deployed. The RoboCupRescue Robot League and the DARPA Robotics Challenge in particular leverage the work of the US Department of Homeland Security (DHS)—US National Institute of Standards and Technology (NIST)—ASTM International (formerly American Society for Testing and Materials) Standard Test Methods for Response Robots project (Jacoff et al. 2014). These test methods, developed in close collaboration with the responder community, reflect elemental decompositions of the capabilities required to answer their operational challenges. This is similar to the basic fitness and skills tests that are decompositions of capabilities required to play games such as soccer and basketball and are used by coaches to shortlist new players.

By constructing competitions around test methods, these research-oriented competitions can direct research towards areas where the currently available solutions are lacking. However, such research is often difficult and resource intensive. This has traditionally only been within the reach of graduate research and higher level undergraduate students at suitably well-resourced institutions. We are leveraging new resources and initiatives in order to allow those with more limited resources, such as secondary school students, early undergraduate students and hobbyists to also access and contribute to such interesting competitions.

Part of this involves providing wider access to the resources required to undertake robotics development and research. Many different initiatives exist in this area, especially with the advent of less expensive 3D printers, robotics kits and electronics for the hobbyist and maker communities.

Lego Mindstorms has been a tremendous catalyst for community development and sharing of robot designs and capabilities, especially among hobbyists and at the primary and secondary school levels. However, flexibility in creating larger, more durable robots is limited. The ability to teach more advanced concepts in design and to disseminate more generally reusable modules is also limited due to the small number of basic components that must be connected together in fairly standard ways.

Another example is ArduPilot (ardupilot.com), an add-on controller for mobile platforms that has become particularly popular among the aerial robotics community. It provides researchers an access to microcontroller or control engineering expertise to integrate working aerial robots at low cost and with flexibility. The Yale OpenHand (Ma et al. 2013), a design for a low-cost hand-type gripper, is backed by a project that seeks to use the open-source community to foster a collection of different designs and variations. Other initiatives, which enable such projects, include the ubiquitous Arduino (arduino.cc) platform and the RepRap (reprap.org) 3D printer, which spawned the majority of the low-cost 3D printers now on the market. Some of particularly powerful characteristics of these initiatives are that the designs that are truly open source, the existence of a vibrant community that allows participants with different levels and types of expertise to share their capabilities, and the relevant licences that encourage commercial entities to reproduce and sell instances of these projects to help bring these developments to a wider audience.

4.6   The Open Academic Robot Kit

Open-source software is a tremendous enabler in robotics research. It allows people around the world to share and build on each other's developments, even if they have never otherwise met, simply by finding software modules that have been published online. They can download and produce fully working software systems much more easily than they could on their own and make modifications and improvements in their area of expertise. Associated usage guides and community-produced documentation allow even newcomers to close the loop on systems and get them working well enough to make contributions in their area of expertise. These can, then, be contributed back to the community. Examples include OpenCV (opencv.org) and the Robot Operating System (ROS) (ros.org).

The OARKit is a community-developed, open-source family of robot designs that aim to replicate this principle for robot hardware. These designs provide researchers and students with working systems that they can improve on, rather than having to reinvent the wheel by developing fully integrated robot systems by themselves.

All of the designs have the following properties that allow anyone, anywhere in the world, to recreate the robot using only the information available on the design's Web page, oarkit.org.

- All non-electronic, non-fastener components are 3D printable, using a sub-$1000 USD printer. The reference designs were printed on a Solidoodle 3 printer, purchased in 2012 for approximately $800 USD.
- All electronic components are readily available off the shelf or via online stores anywhere in the world. Most should be available from a variety of

manufacturers. Some critical parts, such as control boards or smart servos, are not themselves already released under an open licence and only available from one manufacturer. The designs of the OARKit should be easily modifiable to suit a alternative parts from different manufacturers. For example, the reference design uses Dynamixel smart servos but may be easily replaced with HerkuleX smart servos.

- All fasteners are standard M2 or M3 bolts, nuts and washers (or 1/16″ and 1/8″ US equivalents).
- All designs, instructions and source code required for basic demonstration of capabilities are available online, in easily editable form, under an open-source licence (GNU General Public License or Creative Commons Attribution-ShareAlike). This recommendation deliberately allows for their commercial reproduction as long as improvements are contributed back to the community.

These robot designs help researchers and students, especially those with limited mechanical and electronic engineering resources, to construct interesting robots that are a good fit for their requirements. In the past, it would have been necessary for such researchers to adapt off-the-shelf kits, enlist the assistance of mechanical engineering resources or make use of construction kits. Such options often suffer from high cost, long lead times, problematic availability of spare parts, compromised suitability for their specific task or poor reliability, especially where adaptations are designed by those inexperienced in mechanical engineering.

Several high school teams have already made use of the Emu Mini 2 design as shown in Fig. 4.2 and developed their own versions. Some of these, which appeared at the 2016 RoboCupRescue Rapidly Manufactured Robot Competition World Championships in Leipzig, Germany, are shown in Fig. 4.4.

Recent developments in low-cost 3D printing make it possible to produce highly complex parts without needing access to a traditional machine shop. The rapid increase in the availability of low-cost, high-quality and easy to use smart servos, control boards, battery systems, sensors and communications modules also largely eliminates the requirement for expertise in more traditional electronic and lower level control engineering, for students whose interests lie in other aspects of the robotics challenge.

Precision, strength and maximum size of printable components on such low-cost printers will tend to limit the size of robots to the 10-cm–1-m scale with weights up to a few kilograms. We have constructed several reference robot designs that adhere to these principles and can be constructed with a low-cost 3D printer, basic soldering iron and basic hand tools (cutters, pliers, screwdrivers and simple hand drill). The reference designs, shown in Fig. 4.2, are not intended to be the best designs, or even particularly good designs, for any particular purpose. Instead, they are intended to provide starting points that demonstrate the types of robots that may be designed and built using such tools, and that can be easily replicated and extended. For further details please see (Sheh et al. 2014b).

Fig. 4.4 Several of the robots that high school teams have developed, based on the Emu Mini 2 robot shown in Fig. 4.2

4.7 The RoboCupRescue Rapidly Manufactured Robot Competition

Managed properly, robot competitions can be a powerful tool for advancing research. Teams can be highly motivated to solve relevant challenges, extend their capabilities and integrate working systems. Of course, the competition challenges must be properly structured in order to achieve these aims. For over 10 years, we have been running the RoboCupRescue Robot League (Sheh et al. 2012), a research competition aimed at undergraduate research students, Ph.D. students and early career researchers.

This competition has met with considerable success among the research community, in its goals of using the DHS-NIST-ASTM International Standard Test Methods for Response Robots to communicate and disseminate the real-world challenges and capabilities. One high-profile example is the Quince robot

(Nagatani et al. 2011), which was refined over several years of participation in the RoboCupRescue Robot League and proved to be vital in responding to the Fukushima Daiichi Nuclear Power Plant incident.

"The test methods and RoboCupRescue contributed significantly in the process of development of Quince. I can say that Quince would not existed if there were no test methods and no RoboCupRescue. This means we would have had no way of surveillance of 2nd–5th floor of Unit 2 and other units of Fukushima-Daiichi than suicide workers. The cool shutdown would have been much late or impossible in this case.

In addition, we could make appropriate advice to METI (Ministry of Economy, Trade and Industry), TEPCO (Tokyo Electric Power Company) and related companies on the basis of our experience of testing various robots at Disaster City, RoboCupRescue arenas and the NIST standard test field. Various NIST test methods were really valuable. Not only the step field and others for mobility, but also wireless, sensing, human interface, etc., were also valuable."—Professor Satoshi Tadokoro, President, International Rescue System Institute.

While this competition is valuable in promoting and advancing research, the barriers for entry in terms of skills and resources are quite considerable. In an effort to reduce these barriers and to bridge the gap between the RoboCupRescue Robot League (which starts at the final year of undergraduate studies) and the top 10% of teams in the RoboCupJunior Rescue competition (which ends at secondary school), we have developed the RoboCupRescue Rapidly Manufactured Robot Competition (comp.oarkit.org). This is a competition that forms part of, and encapsulates the same challenges that are used in, the RoboCupRescue Robot League competition, at a level that is accessible to high school students.

The existing RoboCupRescue Robot League is held in an arena through which robots must traverse, build maps, find simulated victims and perform other tasks. This is shown in Fig. 4.3. Test method apparatuses, such as the terrains and inspection targets, are placed throughout the arena. The robots must overcome these test method apparatuses in order to traverse the arena and reach the victims and tasks. This scale, with a minimum guaranteed clearance of 1.2 m (the 1.2-m scale), limits robots to a maximum practical width of around 70 cm. This represents the practical scale of robots that can traverse environments that are built to be wheelchair accessible, such as commercial buildings.

Many of the challenges in real response scenarios involve environments that are significantly smaller than this. For example, service passages, ventilation ducting and even occupied dwellings may necessitate robots that can fit through gaps of 30 cm or smaller. Building robots at this smaller scale can introduce additional challenges. However, it can also be easier to work with robots to tackle some new problems at this scale. Robots are often easier to build and handle, cheaper, safer and more forgiving of engineering deficiencies. Their low cost also makes them more suited for deployment as "disposable" robots, which responders are more willing to use in riskier situations knowing that they may not return. The OARKit is intended to operate at this scale.

Hosting the infrastructure to develop and run this competition is also significantly easier at this smaller scale. Instead of teams requiring a dedicated test room for development and the competition requiring a space approximately the size of a small house, a test arena for the Rapidly Manufactured Robot Competition can be set up in as little as 3 m^2. A viable competition arena can be constructed in as little as 6 m^2 using similar techniques to the existing RoboCupJunior Rescue arenas. The current Rapidly Manufactured Robot Competition arena, as deployed at the 2016 International Championships in Leipzig, Germany, is shown in Fig. 4.5.

In combination with the OARKit, the Rapidly Manufactured Robot Competition aims to allow teams of secondary school and undergraduate students to participate in solving the same challenges that the RoboCupRescue Robot League teams do, but at this smaller scale. Some of the challenges are shown in Fig. 4.5. The upper section of the figure shows four challenges laid out separately and are, from left to right:

- Crossing ramps.
- Rotating ramps.
- Continuous ramps.
- Raised ramps.

Fig. 4.5 Rapidly Manufactured Robot Competition arena as configured for the preliminary rounds (*top*) and the finals (*bottom*)

Each of these challenges tests the ability of the robots to traverse different types of rough terrain in a repeatable fashion. Additional challenges exist for capabilities such as manipulation, inspection and precision control.

The same ramps and terrains as exist in the wider RoboCupRescue Robot League competition are scaled down. The same perception and manipulation challenges, and the goals of reaching victims, observing objects of interest and mapping the environment using teleoperated, semi-autonomous and autonomous robots, that exist in the RoboCupRescue Robot League, are maintained. Indeed, from the perspective of the RoboCupRescue Robot League, the Rapidly Manufactured Robot Competition serves as an extension of the existing confined space arena, which simulates the challenges present in a pancake structure collapse (where successive floors have collapsed and produced a 3D labyrinth of passageways through which robots must traverse).

Initial challenges that can be tackled in the Rapidly Manufactured Robot Competition, by students at this level, include:

- Locomotion for small rough terrain robots.
- Camera placement and directed perception for confined space robots.
- Visual display of video and other information on hand-held devices.
- Manipulation and delivery of objects from small robot platforms.
- Sensors for mapping, information recording and situational awareness on small, low-cost robots.

The Rapidly Manufactured Robot Competition arena has been demonstrated at several venues including:

- The RoboCupRescue Robot League World Championships in 2014, 2015 and 2016, most recently held in Leipzig, Germany.
- The 2014 Institute of Electrical and Electronic Engineers (IEEE)—Robotics and Automation Society (RAS) Response Robotics Summer School and Workshop in Perth, Western Australia.
- The 2015 and 2016 RoboCup Junior Western Australia competitions held in Perth.
- The 2016 and 2017 RoboCup Junior North America competition held in New York, USA.
- The 2016 Manufacturing day "Gears on the Gridiron" event held in Maryland, USA.

The first demonstration of the Rapidly Manufactured Robot Competition was run during the 2015 RoboCupRescue Robot League World Championships in Hefei, China, with teams consisting of both secondary school and undergraduate students. The first full competition was held at the 2016 RoboCupRescue Robot League World Championships in Leipzig, Germany, with five teams from Australia, Japan and the USA. Some of their robots appear in Fig. 4.4.

4.8   Summer Schools, Workshops and Camps

While these competitions are a tremendous way to motivate students and researchers to focus their research on the topics of great need, it can be quite difficult for them to actually collaborate and learn from each other during the competition itself. This is especially the case for top teams, who devote their efforts to performing well in the competition and, even with the best intentions, have limited time to disseminate their developments. Many teams will publish their developments as research papers. However, as most researchers have experienced, it can be very difficult to reproduce a result in a paper, especially if the hardware is dissimilar.

Since 2004, we have conducted teaching camps and summer schools in collaboration with the IEEE Robotics and Automation Society (IEEE-RAS) Safety, Security and Rescue Robotics (SSRR) technical committee, the RoboCupRescue Robot League and the DHS-NIST-ASTM International Standard Test Methods for Response Robots project. These events are an avenue through which teams, old and new, share their developments in person. Best-in-class teams from the RoboCupRescue Robot League competition, who have achieved excellence in specific capabilities, join with experts from the wider SSRR community in an event that is half traditional technical sessions and half practical hands-on experimentation. Attendees, who usually range from undergraduate research students to graduate research students and early career researchers, are encouraged to bring their own robots. During the practical sessions, presenters assist them in implementing these capabilities on their own robots. The events are usually coordinated so that attendees eat together and sleep at the same hotel, maximising the time that can be spent discussing and networking.

In recent years, we have experimented with this format to good effect. In 2012, with the support of the IEEE-RAS Members Activity Board's Technical Education Program, we ran the international 2012 IEEE-RAS Safety, Security and Rescue Robotics Summer School in Alanya, Turkey (Sheh and Komsuoglu 2012). For the first time, responders participated in the event alongside the attendees, listening to the lectures and joining in the practicals. For this event, the responders included a bomb squad commander from the USA and leaders from the fire and rescue communities in the USA and Turkey. This innovation proved to be crucial, allowing attendees to fully appreciate the challenges faced by responders in the field and the ways in which their technologies might be used. Such opportunities are ordinarily difficult to come by for most researchers. After all, interested researchers are unlikely to find success in engaging leaders of the responder community by showing up to their local fire or police station. Similarly, responders are likely to have only marginally greater success showing up to the local university research office in the hope of engaging suitable representatives of the international research community.

Providing an environment in which responders and researchers could get to know each other in both a professional as well as a social context was also valuable.

It provided a strong motivator for the academic attendees, who came to understand that their developments can potentially save the lives of people who they now regard as friends. The responders also benefit from the event, as they gain a deeper insight into new and upcoming capabilities and contacts among the research community who share their concerns and can help to answer their questions about new capabilities. This allows them to be better informed as they guide the standard test method development process, work with industry and government, and make recommendations for future procurements of robots.

This concept was expanded in 2013 where, for the first time, such an academic event was co-located with a responder event. The 2013 Response Robotics Summer School (Sheh et al. 2014a) was run jointly with the 2013 Bomb Response Technology Seminar. Co-located at the Maylands Police Complex just outside Perth, Western Australia, and run in collaboration with the Western Australia Police Bomb Response Unit (WAPol BRU), these academic and responder events were unique in that half of the sessions were shared between the two events. Half of these shared sessions were presentations by bomb squad leaders on the challenges and latest capabilities available to the responder. Half were presentations by leaders in the academic robotics community on the latest concepts and developments in relevant technologies. These presentations were followed by joint discussion periods that allowed both groups to better understand each other's challenges and capabilities. These discussion periods also provided a fruitful avenue through which to develop new research ideas that could be immediately applicable.

The 2013 event concluded with two practical demonstrations. First, response robots were tested in standard and prototypical test apparatuses for mobile manipulation that were developed in the DHS-NIST-ASTM International Standard Test Methods for Response Robots project. The test methods were modified for embedding into operational scenarios as part of the "Test Methods in a Suitcase" project and placed throughout a WAPol BRU training facility. Second, academic attendees observed a training deployment of robots to a suspected clandestine drug laboratory. This included driving the robots into the suspected laboratory, collecting samples and taking X-rays of suspicious objects. Senior WAPol BRU personnel were on hand to provide commentary on the events and discuss possible ways in which new and upcoming capabilities in the research community could be adapted, further developed and deployed. Topics that showed promise as a result of these discussions included 3D mapping, human–robot and human–system interfaces, assistive autonomy and alternative and resilient radio communications.

In 2014, once again with the support of the IEEE-RAS Members Activity Board Technical Education Program, we ran the 2014 IEEE-RAS Response Robotics Summer School and Workshop (RRSS+W), also in Perth. This event took the summer school in a somewhat different direction, to engage the secondary education and maker hobbyist communities. It is important to lower the barrier of entry to research in interesting robotic problems and recent developments in resources for building these robots, such as the OARKit, make it easier than ever to do so. By bringing these resources to the secondary education and hobbyist communities, we aim to catalyse further development in these fields. After all, some of the gadgets and widgets that responders find useful, such as novel tool mounts, devices to make entering doors easier and so on, are developed not by traditional manufacturers or by academics but rather by hobbyists and mechanically inclined individuals in workshops and sheds from all around the world.

This event featured the same 50–50 practical development and lecture split as previous events. Best-in-class teams from the RoboCupRescue Robot League, including award-winning teams from Australia, Thailand and Japan, joined with researchers from the USA, Germany and Austria and local responders from the WAPol BRU to disseminate the challenges and best-in-class solutions to the response robotics problem. Talks ranged from challenges in the kinematics of compliant manipulators, recognising large-scale patterns in data and planning for robot learning, to business issues facing robotics start-ups and practical experiences with deploying automation technologies in harsh environments. Demonstrations included novel techniques for mobility with compliant sub-tracks through to a new test method for robot mobility based specifically on the wooden debris, often faced by responders in earthquakes in Japan as shown in Fig. 4.6.

In order to engage the wider maker community and provide the resources for the academic attendees to better explore outside their field, the practical sessions of the summer school and workshop were held at a hackerspace, the Perth Artifactory near Perth in Western Australia (see Fig. 4.6). Attendees included academics at the secondary and tertiary levels from nine countries, and members of the maker community. They participated in the practicals and technical sessions to identify avenues through which initiatives such as the OARKit and the RoboCupRescue Rapidly Manufactured Robot Competition can be used to further encourage secondary school students to pursue science and engineering. The collaboration with this core group of educators will continue with the aim of developing curriculum that bridges the gap between secondary education, undergraduate science and engineering and graduate research. Since this gap has been identified as a focus area by the IEEE-RAS, the developments from this effort will contribute to the wider IEEE-RAS activities in this area through initiatives such as the IEEE-RAS planned pre-college outreach site.

Fig. 4.6 Attendees of the 2014 Response Robotics Summer School and Workshop (*left*) and one of the robots attempting the novel movable debris test method (*right*). Both photographs taken at the Perth Artifactory in Western Australia

4.9 Future Directions

These three interlinked initiatives of the Open Academic Robot Kit, the Rapidly Manufactured Robot Competition and the teaching camps and summer schools, all stem from a solid history. The events held in past decade provide support for the development of technologies required for response robotics and pave the way for future developments in this field. Several groups are currently working to reproduce, improve and extend the designs of the OARKit, including applications in advanced mobility, sensing and manipulation.

In addition, we are working with a core group of secondary school teachers to further develop curriculum in a variety of fields in STEM to encourage students' interests and learning in these fields. There are many possibilities for the design and control of robots at this scale that have been only superficially explored. The OARKit and the Rapidly Manufactured Robot Competition provide the tools and arena with which students, including secondary school level, could make novel contributions.

The ultimate goal is to allow all who can contribute to this vital field—educators, researchers, manufacturers, responders, makers and government—to better work together to advance the state of response robotics and beyond.

References

Jacoff, A., Messina, E., Huang, H.-M., Virts, A., Downs, A., & Norcross, R., et al. (2014). *Guide for evaluating, purchasing, and training with response robots using DHS-NIST-ASTM international standard test methods.* www.nist.gov/el/isd/ks/upload/DHS_NIST_ASTM_Robot_Test_Methods-2.pdf

Ma, R. R., Odhner, L. U., Dollar, A. M. (2013). *A modular, open-source 3D Printed underactuated hand.* In *Proceedings of the 2013 IEEE International Conference on Robotics and Automation (ICRA)*, Karlsrue, Germany.

Nagatani, K., Kiribayashi, S., Okada, Y., Tadokoro, S., Nishimura, T., & Yoshida, T., et al. (2011). Redesign of rescue mobile robot Quince: Toward emergency response to the nuclear accident at Fukushima Daiichi Nuclear Power Station on March 2011. In *Proceedings of the IEEE International Symposium on Safety, Security and Rescue Robotics*, Kyoto, Japan.

Sheh, R., Collidge, B., Lazarescu, M., Komsuoglu, H., & Jacoff, A. (2014a, September). The response robotics summer school 2013: Bringing responders and researchers together to advance response robotics. In *Proceedings of the IEEE/RSJ International Conference on Intelligent Robots and Systems*. Chicago, Illinois, USA.

Sheh, R., Jacoff, A., Virts, A.-M., Kimura, T., Pellenz, J., & Schwertfeger, S., et al. (2012, July). Advancing the state of urban search and rescue robotics through the RoboCupRescue robot league competition. In *Proceedings of the 8th International Conference on Field and Service Robotics (FSR)*. Matsushima, Miyagi, Japan.

Sheh, R., & Komsuoglu, H. (2012, December). The 2012 IEEE robotics and automation society (RAS) safety, security, and rescue robotics (SSRR) summer school. *IEEE Robotics and Automation Magazine*.

Sheh, R., Komsuoglu, H., & Jacoff, A. (2014b, November). The Open Academic Robot Kit: Lowering the barrier of entry for research into response robotics. In *Proceedings of the 12th IEEE International Symposium on Safety, Security and Rescue Robotics*. Lake Toya, Hokkaido, Japan.

Part II
Robotics and STEM Education

Chapter 5
How Have Robots Supported STEM Teaching?

Fabiane Barreto Vavassori Benitti and Newton Spolaôr

Abstract Context: Robotics has assisted teachers to combine technology and engineering topics to concretize science and mathematics concepts in real-world applications. As a result, benefits in different concepts and skills, as well as positive long-term effects, have been observed. Objective: This work aims to identify state-of-the-art robotics applications to support STEM teaching. To do so, we intend to answer six research questions: (1) What concepts are considered and how are they explored? (2) What skills are expected to be developed? (3) How is educational robotics associated with school curriculum? (4) What types of robots are used? (5) What age groups/educational levels are considered? (6) How is educational robotics evaluated? Method: We carried out a systematic literature review to identify, assess, and synthesize relevant papers published from 2013. A protocol developed by us guided the review conduction and enhanced its repeatability with reduced subjectivity. Results: 60 publications able to answer the research questions were summarized. We found that: (1) several STEM concepts have been explored in all educational levels; (2) educational robotics is still frequently associated with teamwork and problem-solving development, extracurricular activities, and LEGO robots; (3) only 25% of the 60 papers quantitatively and qualitatively evaluated learning. Conclusion: Robots support for STEM education has been successful in different scenarios. The inherent flexibility, coupled with the experiences reported by a significant piece of the literature, can inspire new applications of educational robotics.

Keywords LEGO robots · Technology · Review · Methodology · Concepts · Programming

F.B.V. Benitti (✉)
Department of Informatics and Statistics, Federal University of Santa Catarina.
Florianópolis, Santa Catarina, Brazil
e-mail: fabiane.benitti@gmail.com; fabiane.benitti@ufsc.br

N. Spolaôr
Laboratory of Bioinformatics, Western Paraná State University,
Foz do Iguaçu, Paraná, Brazil
e-mail: newtonspolaor@gmail.com

© Springer International Publishing AG 2017
M.S. Khine (ed.), *Robotics in STEM Education*,
DOI 10.1007/978-3-319-57786-9_5

5.1 Introduction

New technological tools are introduced in our life very rapidly. "New iProducts are introduced into the market almost every six months. When watching the Jetsons television program in the 1960s and 1980s, very few people believed that a humanoid robot, such as Rosie, could become a reality in their lifetime. However, robotics in education for school age children has been in existence since the late 1900s" (Eguchi 2014, p. 27).

Robotics, with its multi-disciplinary nature, provides constructive learning environments that are suitable for a better understanding of scientific and non-scientific subjects and it has a significant role on learning Science, Technology, Engineering, and Mathematics (STEM) subjects (Khanlari 2013). Robotics can be especially effective in teaching STEM, as it enables real-world applications of the concepts of engineering and technology and helps to remove the abstractness of science and mathematics. In fact, various robotics activities led to improvements in science, technology, engineering, and/or mathematics learning (Kim et al. 2015).

Robots have the potential to be the next effective add-on to traditional education. The tangibility of robots and the excitements they bring into the classroom environment are considered conducive for learning (Karim et al. 2015). However, the actual contribution of robots in STEM education is not obvious. This brings us to the title of this chapter: How have robots supported STEM teaching?

Although several authors propose to explore the subject (Sect. 5.2), we did not find a recent and systematic study in order to identify state-of-the-art robotics applications to support STEM teaching. Therefore, we carried out a systematic review to find relevant papers published from 2013 to answer six research questions:

1. What concepts are considered and how are they explored?;
2. What skills are expected to be developed?;
3. How is educational robotics associated with school curriculum?;
4. What types of robots are used?;
5. What age groups/educational levels are considered?; and
6. How is educational robotics evaluated?

To do so, we followed the Systematic Literature Review (SLR) protocol indicated in Sect. 5.3. Sections 5.4 and 5.5 discuss the results and describe the conclusions, respectively.

5.2 Background

Our students are digital natives who have grown up using technology. Home computers have been in existence since before they were born—Eguchi (2014) shared that some students thought "B.C." means "before computer"! The world is rapidly changing and educational programs have to adapt to the changes. Thus, this section presents some review studies focused on robotics use as an educational tool.

Potkonjak et al. (2016) claim that the problems that still constrain the full realization of distance education in Science, Technology, and Engineering (STE) lie in the fact that these sciences inevitably require laboratory exercises as part of the skill acquisition process. Thus, the authors summarize the state of the art in virtual laboratories in the fields of STE. Two different points of view to the resolution have appeared. One is to try developing a physical (real) laboratory with distance access, while the other aims to develop a fully software-based virtual laboratory. They argue for the latter option. That paper intends to support wider application of virtual laboratories, and the criteria followed from one crucial requirement, which is operating a virtual laboratory for a student must feel like they are working with real authentic devices in a real authentic space. The authors present a list with 20 virtual laboratory projects classified as follows:

- Two (2) projects in field of the general initiatives which have a wider focus and try to provide a framework for both virtual and remote-access-physical facilities.
- Two (2) projects in field of science-physics.
- Two (2) projects in field of process technology.
- Five (5) projects in field of engineering—non robotic.
- Nine (9) projects in field of robotics.

Sullivan and Heffernan (2016) present a systematic review of research related to the use of robotics construction kits (RCKs) in P-12 learning in the STEM disciplines for typically developing children. The purpose of this review is to answer the question: "How do robotic construction kits function as computational manipulatives in P-12 STEM education?" The synthesis of the literature has resulted in four key insights. First, RCKs have a unique double application: They may be used for direct instruction in robotics (first-order uses) or as analogical tools for learning in other domains (second-order uses). Second, RCKs make possible additional routes to learning through the provision of immediate feedback and the dual modes of unique representation to RCKs. Third, RCKs support a computational thinking learning progression beginning with a lower anchor of sequencing and finishing with a high anchor of systems thinking. And fourth, RCKs support evolving problem-solving abilities along a continuum, ranging from trial and error to heuristic methods associated with robotics study. Furthermore, their synthesis provides insight into the second-order (analogical) uses of RCKs as computational manipulatives in the disciplines of physics and biology.

Can robots in classroom reshape K-12 STEM education and foster new ways of learning? To sketch an answer, Karim et al. (2015) review (no systematic review), side-by-side, existing literature on robot-based learning activities featuring mathematics and physics (see Table 5.1) and existing robot platforms and toolkits suited for classroom environment (36 robots/toolkits were identified). The survey suggests that the use of robots in classroom has indeed moved from purely technology to education, to encompass new didactic fields.

Table 5.1 Summary of the topics covered in educational robotics featuring mathematics and physics (Karim et al. 2015)

Mathematics	Physics
Geometric primitives	Distance, time, and velocity
Counting	Constant speed, acceleration, and deceleration
Multiplication	Work and energy
Decimals	Force, gravity, and friction
Fractions and ratios	Doppler effect
Coordinate system	Fundamentals of electricity
Recognition of quantities	Weight scale and moment computation
Problems with operator	
Graph construction and interpretation	
Angles	

Toh et al. (2016) carried out a systematic review to examine the use of robots in early childhood and lower level education. This paper synthesizes the findings of research studies carried out in the last ten years and looks at the influence of robots on children and education. Four major factors are examined—the type of studies conducted, the influence of robots on children's behavior and development, the perception of stakeholders (parents, children and educators) on educational robots, and finally, the reaction of children on robot design or appearance. The authors point that:

- Robots influence on children's skills development could be grouped into four major categories: cognitive, conceptual, language, and social (collaborative) skills.
- Aside from the main users (children), parents and educators have to be on-board as well in order to increase the chances of success of this kind of programs. Lack of parental support would confine educational robots to applications only inside the classroom.
- Design is usually the last consideration when incorporating robots into an application. However, as studies showed, design could make a difference on robot perception and hence, how the children would interact with it. Unfortunately, little work has been done on this question.

Mubin et al. (2013) present a review on the field of robots in education (post-2000). The aspects reviewed include domain of the learning activity, location of the activity, the role of the robot, types of robots, and types of robotic behavior. The overview shows that robots are primarily used to provide language, science, or technology education and that a robot can take on the role of a tutor, tool, or peer in the learning activity.

Benitti (2012) reviews published scientific literature (until January 2010) on the use of robotics in schools, in order to answer the questions:

1. What topics (subjects) are taught through robotics in schools?;
2. How is student learning evaluated?; and
3. Is robotics an effective tool for teaching? What do the studies show?

The papers reviewed suggest that educational robotics usually acts as an element that enhances learning. However, this is not always the case, as there are studies that have reported situations in which there was no improvement in learning. Thus, Benitti (2012) indicates some factors considered important for an effective use of educational robotics, summarized in what follows:

- The role of the teacher—the teacher plays an important role in stimulating pupils in their school work and giving them positive attitudes, because the teacher has considerable influence over the way in which these tools are received by the pupils.
- There are needs to have a larger space for the pupils to work.
- The working groups should not be too big (maximum 2–3 pupils/Kit).
- The task given to the pupils must be both relevant and realistic to solve.
- Short lessons, tutorials, and debriefings embedded in the problem-solving activities could help students to make the connection between experience and scientific concepts.
- It is important to provide an opportunity for students to explore the robotics kit before requiring them to work on a design challenge.
- Middle-school students, in particular, seem to need relatively specific guidance on how the robotics activities relate to science and engineering.
- The structure of the robotics environment combined with specific pedagogical approaches foster the thinking and science process skills.

5.3 The Systematic Review Process

A SLR (often referred to as a systematic review) is a way of identifying, evaluating, and interpreting all available studies relevant to a particular research question, or topic area. There are many reasons for undertaking a SLR. The most common reasons are (Kitchenham and Charters 2007):

1. To summarize the existence of evidence concerning a treatment or technology.
2. To identify any gaps in current research in order to suggest areas for further investigation.
3. To provide a background in order to appropriately positioning new research activities.

In this work, we applied this systematic approach to provide an overview about recent research in educational robotics, which can assist in the foundation of new researches.

In order to perform this review, we followed a defined process for conducting systematic reviews based on Kitchenham and Charters (2007), covering the stages and activities indicated in Fig. 5.1.

Stage 1: Planning the review

Activity 1.1:
Identification of the
need for a review

Activity 1.2:
Development of a
review protocol

Stage 2: Conducting the review

Activity 2.1:
Identification of
research

Activity 2.2: Selection
of primary studies

Activity 2.3: Study
quality assessment

Activity 2.4: Data
extraction and
monitoring

Activity 2.5: Data
synthesis

Stage 3: Reporting the review

Activity 3.1:
Communicating the
results

Fig. 5.1 Systematic review process: stages and activities

5.3.1 Planning the Review (Stage 1)

Initially, we have performed a search to identify the existence of systematic reviews involving robotics in STEM teaching. The main studies involving review in this issue are listed in Sect. 5.2. We then observed that they do not have a comprehensive and current review on the subject. Thus, we proposed to continue and expand the study carried out by Benitti (2012).

Within the context of this paper, we carried out a systematic literature review in order to examine the state of research in educational robotics, based on the following research questions:

1. What concepts are considered and how are they explored?;
2. What skills are expected to be developed?;
3. How is educational robotics associated with school curriculum?;
4. What types of robots are used?;
5. What age groups/educational levels are considered?; and
6. How is educational robotics evaluated?

To perform the SLR, the databases in which the search should be performed were defined. They were selected based on popularity on previous reviews in education and related areas: ACM, EBSCO, ERIC, IEEE Xplore, Science Direct, Scopus, Springer Link, Web of Science, and Wiley Interscience. Searches were restricted to peer-reviewed articles or conference papers, written in English, and published between 2013 and 2016 (research over the last four years). The search string used was "robotics AND stem."

The selection criteria were used to evaluate each of the studies recovered from the search sources. Thus, the Inclusion Criterion (IC) applied to include relevant studies in our systematic review was:

- IC1: The primary study presents a description of robotics applications to support STEM teaching.

Moreover, the Exclusion Criteria (EC) were used to exclude studies that do not contribute to answer the research questions. In particular, the exclusion criteria considered in our SLR were:

- EC1: The publication does not consider robots application to support the development of STEM concepts or skills.
- EC2: The publication does not evaluate educational robotics in an elementary-, middle-, and high-school context.
- EC3: Duplicated publications by the same authors (Similar title, abstract, results, or text). In this case, only one is kept.
- EC4: Publications composed of only one page (abstract papers), posters, presentations, proceedings, program of scientific events, and tutorial slides.
- EC5: Publications hosted in Web pages which are not accessed through the account of the Federal University of Santa Catarina, the Western Paraná State University, or the University of São Paulo.
- EC6: Publications written in a language different than English.

The second author independently extracted the data—Table 5.2. Conflicts found during this process were resolved by discussion between the authors.

5.3.2 Conducting the Review (Stage 2)

The systematic review was conducted in April 2016 by executing the protocol. The process observed the steps shown in Fig. 5.2, from which 60 out of the 538 pieces of work were selected. The conduction of the studies was done in three steps:

Identification of candidate studies: pieces of work were collected by applying the search string in the databases selected. Table 5.3 outlines the specific search strings used for each source considered.
Selection of relevant studies: using a search string does not guarantee that all the material that was collected is relevant to the research context. Thus, after the identification of publications obtained through the search engines, the studies were analyzed according to the criteria established for exclusion.
Information extraction and synthesis: after setting the final list of relevant publications, the necessary information related to the research objective was extracted from them.

A quality strategy enables one to assess the selected studies in terms of methodological criteria. In Table 5.4, the 8 quality criteria considered by us are mentioned. Note that criteria QC3a, QC4a, and QC5a are designed specifically for selected publications that quantitatively assess learning based on robots, while

Table 5.2 Data extracted from each primary study selected

Group	Information item
Group 1. Publication identification	IE1. Publication ID
	IE2. Publication title
	IE3. Year of publication
	IE4. Authors' name
	IE5. Students' age group/educational level
	IE6. Publication objective
	IE7. Publication source
	IE8. URL
Group 2. Activities reported in the publication	IE9. Duration of the educational robotics activities
	IE10. Motivation to use educational robotics
	IE11. Robot used
	IE12. Robot price
	IE13. Knowledge areas/subjects taught through robotics
	IE14. Skills taught through robotics
	IE15. Teachers' or tutors' training to deal with robots
	IE16. Learning theory used
	IE17. Educational robotics association with curriculum
Group 3. Evaluation described in the publication	IE18. Assessment approach (quantitative, qualitative, or both)
	IE19. Sample size
	IE20. Reliability/validity analysis conducted during quantitative analysis
	IE21. Short description of the quantitative study
	IE22. Statistical test used
	IE23. Study type (non-experimental, quasi-experimental, or experimental)
	IE24. Sample design/target selection conducted
	IE25. Sample composition and coverage
	IE26. Data collection procedure

QC3b, QC4b, and QC5b do the same for pieces of work regarding qualitative evaluations.

One can consider two approaches to deal with the assessment results: (1) supporting the synthesis of the selected publications or (2) specifying more detailed selection criteria. In this work, the first approach was adopted.

Based on the information items extracted and on the quality criteria applied, we conducted a synthesis to answer the research questions.

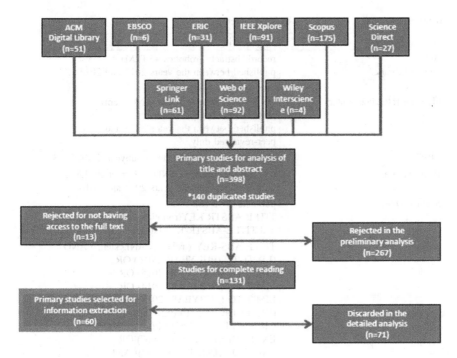

Fig. 5.2 Process of selecting primary studies

5.3.3 Reporting the Review (Stage 3)

Information extracted from each one of the 60 selected pieces of work (Fig. 5.2) is published in a spreadsheet publicly available (https://goo.gl/AJcIyv). By assessing the methodological quality of these publications, it was possible to find 13 papers that accomplishes at least 3 quality criteria (Nugent et al. 2016; Sullivan and Bers 2016; Christensen et al. 2015; Kaloti-Hallak et al. 2015a, b; McKay et al. 2015; Modekurty et al. 2014; Yuen et al. 2014; Abaid et al. 2013; Flannery and Bers 2013; Nag et al. 2013; Sullivan and Bers 2013; Kazakoff et al. 2013).

Section 4 focuses on these highlighted papers and on the synthesis conducted by us to answer the research questions.

5.4 Results and Discussion

In what follows, the research questions are answered according to findings from the 60 papers selected by the systematic review. Table 5.5 specifies an identifier for each of these publications. Additional information regarding the papers is available in the supplementary material (https://goo.gl/AJcIyv).

Table 5.3 Digital libraries and associated search strings

Database	Search string
ACM Digital Library	recordAbstract:(+robotics +STEM), published between the years 2013 and 2016
EBSCO (Teacher reference center)	title (robotics AND stem) OR subject terms (robotics AND stem) OR abstract (robotics AND stem), published between the years 2013 and 2016 peer-reviewed only
ERIC	abstract:(robotics AND stem) pubyear: 2016-2013
IEEE Xplore	(("Abstract":robotics) AND "Abstract":STEM), published between the years 2013 and 2016
Science Direct	pub-date 2012 and TITLE-ABSTR-KEY(robotic) and TITLE-ABSTR-KEY(stem).
Scopus	TITLE-ABS-KEY (robotics AND stem) AND (LIMIT-TO (PUBYEAR, 2016) OR LIMIT-TO (PUBYEAR, 2015) OR LIMIT-TO (PUBYEAR, 2014) OR LIMIT-TO (PUBYEAR, 2013)) AND (EXCLUDE (DOCTYPE, "ch") OR EXCLUDE (DOCTYPE , "ed") OR EXCLUDE (DOCTYPE, "no") OR EXCLUDE (DOCTYPE, "sh")) AND (LIMIT-TO (SUBJAREA, "ENGI") OR LIMIT-TO (SUBJAREA, "COMP") OR LIMIT-TO (SUBJAREA, "SOCI") OR LIMIT-TO (SUBJAREA, "EART") OR LIMIT-TO (SUBJAREA, "MATH") OR LIMIT-TO (SUBJAREA, "AGRI") OR LIMIT-TO (SUBJAREA, "MATE") OR LIMIT-TO (SUBJAREA, "CENG") OR LIMIT-TO (SUBJAREA, "NEUR") OR LIMIT-TO (SUBJAREA, "ENVI") OR LIMIT-TO (SUBJAREA, "DECI") OR LIMIT-TO (SUBJAREA, "MULT") OR LIMIT-TO (SUBJAREA, "PSYC"))
Springer Link	(robotic AND stem), within Education & Language Social Sciences, published between the years 2013 and 2016 (robotic AND stem), within Education & Language Computer Science, published between the years 2013 and 2016
Web of Science	(robotics AND stem), years: 2013–2016 and type: article
Wiley Interscience	(robotics AND stem) in Article Titles OR (robotics AND stem) in Abstract OR (robotics AND stem) in Keywords between the years 2013 and 2016

Table 5.4 Criteria used to access the methodological quality of selected publications

ID	Quality criterion
QC1	Have the teachers or the mentors been trained to use educational robotics?
QC2	Is educational robotics application based on any learning theory?
QC3a	Is there a comparison or control group? (Greenhalgh 2000)
QC4a	Does the quantitative assessment described in the publication involve a statistical analysis of significance? (Crombie 1996)
QC5a	Is any reliability or validity analysis carried out during the quantitative analysis?
QC3b	How well defined are the sample design/target selection of cases/documents? (Petticrew and Roberts 2005)
QC4b	How well is the eventual sample composition and coverage described? (Petticrew and Roberts 2005)
QC5b	How well was qualitative data collection carried out? (Petticrew and Roberts 2005)

5.4.1 What Concepts Are Considered and How Are They Explored?

We tried to pull of the selected studies which contents were the focus of learning through robotics. After, we classify the content in STEM areas. This was an arduous task, since few publications clearly explain the contents that were addressed. For this reason, some contents are very specific and other contents were described more generally (such as robotics). Tables 5.6 and 5.7 indicate subjects considered in the 60 publications collected from the literature.

By focusing on the 13 papers highlighted on the previous quality assessment (Sect. 5.3), one can find examples of how the concepts are explored. Nugent et al. (2016), for example, report positive results achieved by a comprehensive program for the youth conducted on informal (out of school) learning environments, such as robotics camps, clubs, and competitions. This program requires from students science, engineering, mathematics, and robotics concepts by including in each activity:

- Introductory material related to the focused concept and skills.
- A guided primary exercise with step-by-step instructions.
- An exercise that asks the youth to either research applications of the concepts or to record their efforts.
- A team exercise based on robots.
- A challenge that asks the youth to work as a team to solve a given problem with little facilitator guidance.

Their program is associated with a curriculum that consists of nearly 40 h of instruction, in which each task typically needing one to four hours to be completed. Samples of tasks cover such skills such as writing simple programs, programming the movement of robot motors, and the navigation based on sensors. Most of the robotics tasks are accomplished by pairs of students, while more advanced challenges are solved by groups of three or four students.

Table 5.5 Identifiers of the 60 papers selected by the systematic review

Identifier	Paper citation	Identifier	Paper citation
#1	Abaid et al. (2013)	#31	Laut et al. (2015)
#2	Akiva et al. (2015)	#32	Liu et al. (2013)
#3	Ayar (2015)	#33	Martin et al. (2013)
#4	Barger and Boyette (2015)	#34	McKay et al. (2015)
#5	Brown and Howard (2014)	#35	McKay et al. (2013)
#6	Bussi and Baccaglini-Frank (2015)	#36	Modekurty et al. (2014)
#7	Cateté et al. (2014)	#37	Montironi et al. (2015)
#8	Chen et al. (2015)	#38	Nag et al. (2013)
#9	Christensen et al. (2015)	#39	Nemiro et al. (2015)
#10	Chung et al. (2014b)	#40	Nugent et al. (2016)
#11	Chung (2014)	#41	Phamduy et al. (2015)
#12	Chung et al. (2014a)	#42	Pinzon and Huerta (2014)
#13	de Cristóforis et al. (2013)	#43	Prayaga et al. (2013)
#14	Deken et al. (2013)	#44	Qidwai et al. (2013)
#15	Eguchi (2016)	#45	Rao (2015)
#16	Erickson-Ludwig (2015)	#46	Rubenstein et al. (2015)
#17	Flannery and Bers (2013)	#47	Sahin et al. (2014)
#18	Galley et al. (2015)	#48	Sala et al. (2014)
#19	Garcia et al. (2014)	#49	Saleiro et al. (2013)
#20	Gucwa and Cheng (2014)	#50	Sallee and Peek (2014)
#21	Hamner and Cross (2013)	#51	Senaratne et al. (2014)
#22	He et al. (2015)	#52	Suescun-Florez et al. (2013)
#23	He et al. (2014)	#53	Sullivan and Bers (2013)
#24	Jackson (2013)	#54	Sullivan and Bers (2016)
#25	Jeon et al. (2016)	#55	Talley et al. (2013)
#26	Kaloti-Hallak et al. (2015a)	#56	Tewolde and Kwon (2014)
#27	Kaloti-Hallak et al. (2015b)	#57	Tuluri (2015)
#28	Karp and Maloney (2013)	#58	Ucgul and Cagiltay (2014)
#29	Kazakoff et al. (2013)	#59	van Delden and Yang (2014)
#30	Larkins et al. (2013)	#60	Yuen et al. (2014)

The reported results suggest that the program promotes gain in knowledge in some areas, such as engineering and robotics. This finding, based on non-experimental and quasi-experimental studies, may reflect the lack of an engineering course in middle-school and the unique abilities required to program a robot.

Sullivan and Bers (2016) assess an 8-week robotics curriculum that supports teaching foundational robotics for children from prekindergarten to second-grade classes. The program consists of the following lessons, which guide children to explore basic robots parts, sensors, and robot navigation.

Table 5.6 Subjects and related topics explored in the selected papers: science and technology

Science	Technology
Hypothesis formulation (#13)	Robotics (#1, 2, 3, 4, 8, 10, 13, 14, 15, 16, 18, 21, 24, 27, 31,
Astronomy (#42)	36, 38, 39, 44, 45, 46, 50, 53, 58, 60)
– Terraforming (#30)	Art platforms (#54)
– Space exploration (#30)	Computer aided engineering (#48)
– Satellite control (#38)	Computer science (#7)
Body resistance (#57)	– Cyber security (#18)
Buoyancy (#34, 35)	– Input-output (#26)
Ecology	– Interfacing with sensors (#26)
– Recycling (#29, 49)	Energy monitoring (#9)
– Resource reuse (#49)	Excel and matlab (#48)
– Natural environment (#1, 31)	Gear (#10, 34, 35, 37, 40)
– Biodiversity (#31)	Navigation (#44)
– Marine pollution (#41)	Programming (#2, 3, 6, 13, 16, 17, 18, 23, 24, 25, 30, 32, 33, 34,
Electrical conductivity (#57)	35, 36, 37, 38, 39, 40, 44, 45, 46, 47, 52, 55,58, 59, 60)
Inquiry (#40)	– Visual programming (#29, 51, 53)
Investigation (#28)	– Control flow (#53, 54)
Marine science (#1, 31)	– Programming logic (#12, 54)
Materials science (#48)	– Java programming (#12)
Newton's law of cooling (#57)	– Programming action sequencing (#54)
Photonics and lasers (#48)	– Smart phone programming (#56)
Physics concepts (#43)	– Algorithmic thinking (#32)
Thermal science (#48)	– General programming (#32)
	Sensors (#4, 10, 13, 18, 30, 40, 44, 46, 53, 54, 58, 60)
	Telemetry (#38)
	Sturdy building (#54)

- What is a robot and what is programming?;
- What is a sound sensor?;
- What are repeat loops?;
- What are distance and light sensors?;
- What are conditional statements?; and
- Final project.

First, the children explored the basic parts of a robot and had the first contact with a programming environment based on wooden blocks with barcodes. After constructing the robot, organizing the blocks, and using an embedded scanner, the students read the barcodes associate with commands to program the robot to dance. In the next lessons, the children made programs to allow the robot to interact with

Table 5.7 Subjects and related topics explored in the selected papers: engineering, mathematics, and others

Engineering	Mathematics	Others
Advanced manufacturing (#48)	Algebra (#4, 20)	"Me and My Community" (#54)
Caudal fin building (#1, 31)	Fractions (#40)	Theater and Poetry (#21)
Computer engineering (#48)	Functions (#10)	Arts (#25)
Design (#4, 18, 30, 37, 38, 44, 55)	Geometric progress (#12)	Music (#12)
Electrical engineering (#48)	Geometry (#6, 10, 37)	Boat structure and navigation (#8)
Electromagnets (#59)	Math education (#5)	
Electronics (#3, 16, 18)	Ratios (#40)	
Engineering design (#21, 40)	Reasoning (#49)	
Engineering design process (#34)	Sequencing (#29)	
Fluid power system (#19)	System of equations (#10)	
Geotechnics (#52)	Math education (#5)	
Hydraulics (#19)		
Industrial robotic principles (#59)		
Manufacturing (#3, 4)		
Mecatronics (#31)		
Mechanics (#4, 24, 30, 37, 58)		
Mechanics of materials (#48)		
Mobile robotics (#56)		
Pneumatics (#19)		
Solar energy use (#55)		
Some basic engineering (#59)		
Robot modeling (#47)		

humans by the sound sensor and with the environment with the other sensors. Finally, the pupils made floor maps and programmed the robots to navigate on it.

As part of the study assessment, the children were asked to help the researcher identify different parts of the robot and their functions. As a result, it was noted that the students had a good understanding of the functions of each robot part. In addition, no significant difference was found among the classes, indicating that all children were able to master the robotics concepts similarly, regardless of what grade they were in.

It should be emphasized that each classroom moved at a pace that was comfortable for it. Thus, although all classes conducted robotics and programming activities during 8 weeks, not all grades followed all topics. In particular, prekindergarten children focused on the initial lessons, while the remaining classes were able to spend time experimenting with the different sensors and programming constructs.

McKay et al. (2015) use WaterBotics, a challenge-based curriculum, to stimulate k-12 students to develop physical science concepts, such as buoyancy and stability. These concepts are explored in underwater robots, demanding a complexity level not found in many land-based robotics programs. To do so, a series of four challenges (missions) is managed to gradually lead to the production of a fully functional robot.

Rescue a drowning swimmer represented by ping pong balls by programming a single motor to follow a straight trajectory.

Clean a pollution spill (scattered balls) by programming two motors to enable steering and 2D movement.

Disable underwater mines by using a third motor to dive to the bottom of the pool in a 3D movement to achieve inverted plastic cups.

Collect objects (balls) from an imaginary sunken ship and deposit them in bins by using a fourth motor to grab and release the objects.

In particular, the youth focus on a group of robot capabilities in each mission, planning, designing, building, testing, and iteratively improving a robot. This allows students to benefit from knowledge and experience gained in previous missions. Although the method evaluation did not show significant improvement in concept learning, students felt they had learned and their teachers agreed.

Flannery and Bers (2013) consider the CHERP programming language to support children to explore powerful ideas from technology-based domains that are often and unnecessarily reserved for older children or adults. In particular, their study is conducted as part of the TangibleK project, a program intended to detail what kindergarteners can understand about the robots programming concept. As part of the study, the participants are exposed to programming concepts and challenges, while the reasoning of each child is categorized into different developmental levels. In particular, the students attended a session for preassessments and introduction to the technologies. Afterward, they participated in three sessions in which they built a robot vehicle, learned new programming concepts, attempted a challenge to program the robot to dance, and reflected on their work.

By analyzing the programming achievement, the authors expect that children in different stages of cognitive development would benefit from learning goals, activities, and scaffolding designed specifically for their characteristics.

It should be emphasized that educational robotics supported the exploration of different concepts in other papers found by the systematic review, as illustrated by Akiva et al. (2015), Ayar (2015), Barger and Boyette (2015), Erickson-Ludwig (2015), McKay et al. (2015), Montironi et al. (2015), Rubenstein et al. (2015), Brown and Howard (2014), Chung et al. (2014b), Garcia et al. (2014), Laut et al. (2015), Karp and Maloney (2013), Larkins et al. (2013), Liu et al. (2013), Martin et al. (2013), Suescun-Florez et al. (2013)—Tables 5.6 and 5.7.

5.4.2 What Skills Are Expected to Be Developed?

The most usual skills found in the 60 reviewed papers are related to the following:

Teamwork and
Problem solving.

In addition, some selected publications report experiences on competition, mathematical skills, communication, brainstorming, presentation, creative thinking, critical thinking, strategy making, and leadership.

Nag et al. (2013) introduce the use of collaborative games as a bridge between space-based engineering and STEM and computer science education. To develop the teamwork, strategy-making, leadership, and communication skills, the authors assume as a tenet that collaboration and competition are not mutually exclusive. In particular, the potential to establish social interaction in some games is noted as an issue to support collaboration, even in a competition environment. As part of the evaluation of the learning quality, it was recorded that the k-12 participants found their leadership, teamwork, and strategy-making skills the most improved. For the leadership skill, this finding was similar to the one reported by the students' mentors.

Although Kazakoff et al. (2013) applied the CHERP programming language for children, their focus was on developing sequencing. This important mathematical skill for early childhood is a component of planning and involves putting objects or actions in the correct order (Zelazo et al. 1997). As a result of using robots in this context, it was found a significant increase in terms of sequencing scores for both prekindergarten and kindergarten students.

5.4.3 How Is Educational Robotics Associated with School Curriculum?

Figure 5.3 summarizes how educational robotics is associated with school curriculum. In particular, three categories are considered:

Curricular: papers using robots according to a curriculum in a school.
Extracurricular: robots applications unrelated to a school curriculum.
Hybrid: publications that either combine out-of-school or afterschool activities with a robotics curriculum, or report in-school and out-of-school activities grounded on a curriculum.

The systematic review results suggest that most of the selected papers fall into the extracurricular or hybrid categories. In what follows, a few of the 13 publications highlighted in the previous assessment are considered.

Sullivan and Bers (2013) follow the TangibleK robotics program, which consists of six lessons regarding topics related to robotics, physics, and programming.

Fig. 5.3 Types of association
between educational robotics
and a curriculum

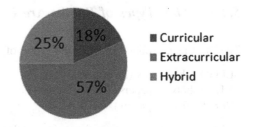

In particular, the lessons include introduction to a design process, robotics kits and control flow in a program and sensors. Furthermore, the curriculum is designed for a minimum of 20 h of classroom work and is implemented with the support of the CHERP language and the LEGO Mindstorms robot. The authors report an evaluation of TangibleK in 3 kindergarten classrooms, such that 53 students were exposed to the experience. It should be emphasized that the program was also considered in other publications, as exemplified in Flannery and Bers (2013), Kazakoff et al. (2013).

McKay et al. (2015) exemplify the hybrid category. In particular, the WaterBotics curriculum was carried out in in-school and out-of-school environments. This program includes science lessons embedded within missions with "achievements" that teams can earn. In this sense, the program is an alternative to competition-driven curricula, making it attractive to youth who may not yet have established STEM identities, interest, and self-confidence. Moreover, WaterBotics focus on a few instructional design principles:

Design-based activities support science learning.
Robotics learning represents a powerful learning opportunity for diverse youth.
Science content learning is scaffolded through mastery of a series of increasingly complex design challenges.

Abaid et al. (2013) illustrate the extracurricular category with an outreach experience in the New York Aquarium. It was designed to ignite K-12 students' interest in STEM and attract them toward engineering careers. As an educational tool, the authors considered a robotic fish easy to control by young participants. To act as a bridge between the knowledge of elementary- and middle-school students and the authors, two high-school students were selected and trained. In this scenario, a student exposed to the experience typically participated in the following route:

Observing different fish species.
Making of caudal fins.
Mounting a fin on the robot and controlling its swimming.
Attending to a high-school students' talk.
Observing robot pieces.
Answering a survey about the experience.

5.4.4   What Types of Robots Are Used?

The most usual robots found in the group of 60 chosen papers correspond to:

LEGO robotic kits (25 papers).
VEX robots (3 papers).
Robotic fish (3 papers).

LEGO robots consists in the most popular choice, as indicated in 25 out of the 60 papers. The potential to include varied sensors and the support for a few traditional programming languages are some reasons that explain this popularity. They are used to support the development of different concepts and skills, as illustrated by Nugent et al. (2016), Kaloti-Hallak et al. (2015a, b), McKay et al. (2015), Yuen et al. (2014), Flannery and Bers (2013), Kazakoff et al. (2013), Sullivan and Bers (2013).

VEX robots, in turn, are selected for educational robotics in He et al. (2014, 2015), Liu et al. (2013). These machines can be combined with remote control devices and microcontrollers associated with an environment that supports visual programming. Benefits for VEX users include robustness against shocks and good availability of sensors.

As shown in the literature, robotic fish is also considered in a few papers (Laut et al. 2015; Phamduy et al. 2015; Abaid et al. 2013). This bioinspired robot contains an artificial flapping tail that approximates the locomotion of some animal species and can be managed remotely by an interface similar to a video game controller. The authors indicate that the entire system costs under US$100 on a limited production basis.

The following robot models are employed by only one or two publications: Kiwi (Sullivan and Bers 2016), Linkbot (Montironi et al. 2015; Modekurty et al. 2014), Aerobot (Rubenstein et al. 2015), Hummingbird (Akiva et al. 2015), Proteus (Ayar 2015), Bee-bot (Bussi and Baccaglini-Frank 2015), a boat robot (Chen et al. 2015), Darwin-OP (Brown and Howard 2014), SIFEB (Senaratne et al. 2014), Khepera (de Cristóforis et al. 2013) and Infante (Saleiro et al. 2013). One also can find alternatives in the literature, such the use of virtual robots (Gucwa and Cheng 2014; Martin et al. 2013) or telerobotics (Prayaga et al. 2013). The latter idea is particularly attracting when the cost per student must be reduced, while the participants are still able to control real robots by a Web interface.

5.4.5   What Age Groups/Educational Levels Are
            Considered?

It was found in the selected papers the use of robots to support teaching in different educational levels. Table 5.8 associates each level with the corresponding publications, which in turn are indicated by the identifiers specified in Table 5.5.

Christensen et al. (2015) evaluate their program with three groups of participants:

Middle Schoolers Out to Save the World (MSOSW).
Communication, Science, Technology, Engineering and Mathematics (CSTEM) program.
Texas Academy of Mathematics and Science (TAMS) program.

The MSOSW project involved participants from 8 US states. Supervised by teachers, they monitor energy use in specific places and study natural environment topics. The CSTEM afterschool program, in turn, engages middle-school students in multi-age groups to solve challenges by industry professionals and learning activities regarding geoscience, creative writing, sculpture, film, and photography. Finally, the TAMS residential program is designed for high-school juniors and seniors who are high achievers and interested in mathematics and science.

Sullivan and Bers (2016) report learning outcomes achieved in a US public school that serves children in prekindergarten through 3rd grade. A similar school type was found in Kazakoff et al. (2013). It should be emphasized that some authors also used alternative terms to refer to the considered educational levels, such as K-4, K-8, and K-12 (Yuen et al. 2014; Nag et al. 2013; Suescun-Florez et al. 2013).

5.4.6 How Is Educational Robotics Evaluated?

The following types of learning assessment were identified in the 60 papers: quantitative, qualitative, or both. Figure 5.4 indicates the frequency of each type in the group of publications.

Table 5.8 Educational levels found in the selected papers

Learning theories	References
Authentic learning	#39
Cognition-based theories	#17, 30
Collaborative learning	#22, 47, 60
Constructivist/constructionism	#13, 16, 29, 53, 58
Elaboration theory	#58
Embodied theory	#33
Experiential learning	#2, 4, 40, 50
Interactive learning	#25
Meaningful learning	#26
Problem-based learning	#19, 34
Project-based learning	#8, 36, 52
Self-determination theory	#3, 27
Semiotic mediation	#6

As can be seen, quantitative assessments are the most common choice (55%). McKay et al. (2015) illustrate this category by taking surveys and students' achievements for further analysis. In particular, surveys intended to capture students' interest, enjoyment, and learning, while the achievements attempted to identify learning outcomes in physics. They also conducted the Mann-Whitney U test to find whether differences in specific results were significant. As an alternative to surveys, Yuen et al. (2014) applied a group observation form to capture information related to students' observable behaviors and interactions that occur in collaboration during robotics projects.

Qualitative evaluations are found in 20% of the 60 references. To collect data from children, Ucgul and Cagiltay (2014) considered multiple strategies, such as semi-structured interviews, participant-observation, field notes, and surveys. In addition, they applied intercoder agreement (Creswell 2007), triangulation (Merriam 2009), and other approaches to ensure the trustworthiness of the study. Nemiro et al. (2015) also verified intercoder agreement in their evaluation in elementary schools.

Both quantitative and qualitative assessments were identified in some publications. Abaid et al. (2013), for example, apply surveys with open-ended questions and with questions answered according to a Likert scale to participants of an outreach program associated with a US aquarium. Modekurty et al. (2014) and Kaloti-Hallak et al. (2015a) also illustrate this hybrid category.

Besides the assessment type, the research design was collected from the 60 pieces of work. Although the non-experimental setting is predominant (53 out of 60), there are a few illustrations of quasi-experimental settings. Nugent et al. (2016), for example, compared outcomes found in robotics summer camp with the ones achieved by control group composed of students identified by some educational service units as youth with interest in technology and robotics. The experiment showed that the camp intervention led to a learning improvement, supplementing the non-experimental setting results.

Fig. 5.4 Type of assessment conducted in a selected paper

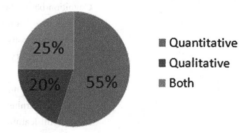

5.4.7 Other Findings

Besides the issues reported previously, we also collected additional information from the 60 references selected in April 2016. The publication year was one of the findings. As Fig. 5.5 shows, a regular number of relevant papers have been published. In addition, it is expected that the number of pieces of work from 2016 increases as the year goes by.

Another finding from the systematic review is that 26 educational robotics applications are clearly grounded in learning theories. Table 5.9 shows 27 study identifiers, as Ucgul and Cagiltay (2014) consider two theories in their work. The learning theory issue can be relevant, as represents a link between practice and theory in teaching supported by robots. Figure 5.6 shows that, from the theories used in more than 2 papers, the constructivist/constructionism theory (Papert 1980; Piaget and Inhelder 1967) is the most frequent one (5 occurrences), while experiential learning comes next (4 occurrences).

Fig. 5.5 Number of pieces of work published per year

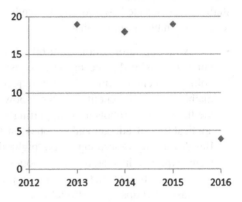

Table 5.9 Learning theories used in the literature

Learning theories	References
Authentic learning	#39
Cognition-based theories	#17, 30
Collaborative learning	#22, 47, 60
Constructivist/constructionism	#13, 16, 29, 53, 58
Elaboration theory	#58
Embodied theory	#33
Experiential learning	#2, 4, 40, 50
Interactive learning	#25
Meaningful learning	#26
Problem-based learning	#19, 34
Project-based learning	#8, 36, 52
Self-determination theory	#3, 27
Semiotic mediation	#6

Fig. 5.6 Learning theory most used in the selected references

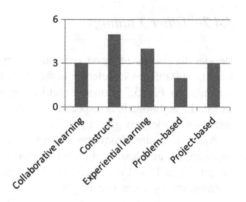

5.5 Conclusions and Perspectives

The present systematic review about robotics in STEM teaching was performed through the elaboration of a predefined protocol review that allowed us to identify and select our primary studies. From the 538 pieces of work initially identified, 60 studies were selected. Based on the synthesis conducted as part of the systematic review, we have observed that:

- By extracting information from 60 relevant publications, we noted that a large number of STEM concepts were considered in the literature. Although technology and engineering are more frequently associated with robots, science and mathematics also benefit from these powerful machines. These findings indicate the flexibility of robots as a supporting tool for learning. It also can inspire new applications of educational robotics on the same concepts or in related ones. Besides briefly describing a few highlighted examples of how these concepts were explored in classrooms, robotics camps, and competitions, the current review reports information on this topic for all the 60 references by using a spreadsheet (https://goo.gl/AJcIyv).
- The mostly observed skills remain the same as reported in previous studies: teamwork and problem solving. We also can highlight the possibility of exploring the engineering design process, illustrated by papers such as McKay et al. (2015). When defining the problem, planning solutions, making a model, testing the model, and reflecting and redesigning robots, students not only learn how technology works, but they also apply the skills and content knowledge learned in a meaningful way.
- The use of robotics as a predominantly extracurricular activity remains. It was not aim of this study to understand why—however, it can be a good topic for future research. Some hypotheses that can be considered are as follows:
  - The schools do not have infrastructure to meet the amount of regular students.
  - Teachers have no knowledge to incorporate robotics into their practice.
  - It is not feasible due to the number of students and the need for follow-up.

– Lack of methodological support.
– Schools are not convinced that the result will be positive.

Anyway, we need to understand the context, the needs, and then propose solutions that integrate more robotics in the practice of the classroom.

- Karim et al. (2015) identified 36 robots/toolkits suitable for use in the classroom. However, as noted in Benitti (2012) and Karim et al. (2015), LEGO robots still consists in the most popular choice.
- We found robotics research at all educational levels. This is good news. However, kindergarten and prekindergarten concentrate the smallest number of studies. We believe that there is much to research at this level.
- "Each primary type of qualitative data contributes unique and valuable perspectives about student learning to the outcomes-based assessment process. When used in combination, a more complete or holistic picture of student learning is created" (Bresciani et al. 2009, p. 61). Similarly, we believe that the evaluation of research involving the application of robotics should include quantitative and qualitative analysis. At this point, only 25% of the considered studies used both approaches. Moreover, we highlight the need for more experimental studies in this area, once non-experimental settings are predominant. We agree with Karim et al. (2015) and believe that it is necessary to standardize evaluation techniques used to quantify robot-based learning. In addition, statistical analysis, surveys, and interviews could be merged to provide more complete findings.

It should be emphasized that this study was based on 60 articles located by using specific search criteria in 9 bibliographic databases. Other criteria and databases would, perhaps, have yielded more studies. The study should, therefore, be considered an attempt to explore the potential of educational robotics in STEM area, rather than a complete overview.

References

Abaid, N., Kopman, V., & Porfiri, M. (2013). An attraction toward engineering careers: The story of a Brooklyn outreach program for KuFFFD12 students. *IEEE Robotics Automation Magazine, 20*(2), 31–39. doi:10.1109/MRA.2012.2184672

Akiva, T., Povis, K. T., & Martinez, A. (2015). Bringing in the tech: using outside expertise to enhance technology learning in youth programs. *Afterschool Matters, 22*, 45–53.

Ayar M. C. (2015). First-hand experience with engineering design and career interest in engineering: An informal STEM education case study. *Educational Sciences: Theory & Practice, 15*(6), 1655–1675. doi:10.12738/estp.2015.6.0134

Barger, M., & Boyette, M. A. (2015). Do k-12 robotics activities lead to engineering and technology career choices? In *American Society for Engineering Education Annual Conference and Exposition* (pp. 1–9). doi:10.18260/p.23895

Benitti, F. B. V. (2012). Exploring the educational potential of robotics in schools: A systematic review. *Computers & Education, 58*, 978–988. doi:10.1016/j.compedu.2011.10.006

Bresciani, M., Gardner, M., & Hickmott, J. (2009). *Assessment methods in the book: Demonstrating student success, a practical guide to outcomes-based assessment of learning and development in student affairs*. United States: Stylus Publishing, LLC, Sterling

Brown, L. N., & Howard, A. M. (2014). The positive effects of verbal encouragement in mathematics education using a social robot. In *IEEE Integrated STEM Education Conference* (pp. 1–5). doi:10.1109/ISECon.2014.6891009

Bussi, M. G. B., & Baccaglini-Frank, A. (2015). Geometry in early years: Sowing seeds for a mathematical definition of squares and rectangles. *ZDM Mathematics Education, 47*(3), 391–405. doi:10.1007/s11858-014-0636-5

Cateté, V., Wassell, K., & Barnes, T. (2014). Use and development of entertainment technologies in after school STEM program. In *ACM Technical Symposium on Computer Science Education* (pp. 163–168). doi:10.1145/2538862.2538952

Chen, Y. K., Chang, C. C., & Tseng, K. H. (2015). The instructional design of integrative STEM curriculum: A pilot study in a robotics summer camp. In: *International Conference on Interactive Collaborative Learning* (pp. 871–875).

Christensen, R., Knezek, G., & Tyler-Wood, T. (2015). Alignment of hands-on STEM engagement activities with positive STEM dispositions in secondary school students. *Journal of Science Education and Technology, 24*(6), 898–909. doi:10.1007/s10956-015-9572-6

Chung, C. J. C. J. (2014). Integrated STEAM education through global robotics art festival (GRAF). In *IEEE Integrated STEM Education Conference* (pp. 1–6). doi:10.1109/ISECon.2014.6891011

Chung, C. J. C. J., Cartwright, C., & Chung, C. (2014a). Robot music camp 2013: An experiment to promote STEM and computer science. In *IEEE Integrated STEM Education Conference* (pp. 1–7). doi:10.1109/ISECon.2014.6891012

Chung, C. J. C. J., Cartwright, C., & Cole, M. (2014b). Assessing the impact of an autonomous robotics competition for STEM education. *Journal of STEM education, 15*(2), 24–34.

Creswell, J. W. (2007). *Qualitative inquiry and research design: Choosing among five traditions*. Thousand Oaks, United States: Sage.

de Cristóforis, P., Pedre, S., Nitsche, M. A., Fischer, T., Pessacg, F., & Pietro, C. D. (2013). A behavior-based approach for educational robotics activities. *IEEE Transactions on Education, 56*, 61–66.

Crombie, I. K. (1996). *The pocket guide to appraisal*. London, United Kingdom: BMJ Books.

Deken, B., Koch, D., & Dudley, J. (2013). Establishing a robotics competition in an underserved region: Initial impacts on interest in technology and engineering. *Journal of Technology, Management & Applied Engineering, 29*(3), 1–9.

van Delden, S., & Yang, K. P. (2014). Robotics summer camps as a recruiting tool: A case study. *Journal of Computing Sciences in Colleges, 29*(5), 14–22.

Eguchi, A. (2014). Robotics as a learning tool for educational transformation. In *International Workshop Teaching Robotics, Teaching with Robotics & International Conference Robotics in Education* (pp. 27–34).

Eguchi, A. (2016). RoboCupjunior for promoting STEM education, 21st century skills, and technological advancement through robotics competition. *Robotics and Autonomous Systems, 75*, 692–699. doi:10.1016/j.robot.2015.05.013

Erickson-Ludwig, A. (2015). A college lead informal learning engineering education program for school aged youth. In *IEEE Integrated STEM Education Conference* (pp. 83–87). doi:10.1109/ISECon.2015.7119951

Flannery, L. P., & Bers, M. U. (2013). Let's dance the "robot hokey-pokey!": Children's programming approaches and achievement throughout early cognitive development. *Journal of Research on Technology in Education, 46*(1), 81–101.

Galley, D., Martin, G. S., Stone, J. C., Hunt, B., Laswell, J., Mortensen, L., et al. (2015). A view from the high school/two year college partnership interface: Our best practices employed in engineering and technology education. In *American Society for Engineering Education Annual Conference and Exposition* (pp. 1–27).

Garcia, J. M., Kuleshov, Y. A., & Lumkes, J. H. (2014). Using fluid power workshops to increase STEM interest in k-12 students. In *American Society for Engineering Education Annual Conference and Exposition* (pp. 1–12).

Greenhalgh, T. (2000). *How to read a paper: The basics of evidence-based medicine*. London, United Kingdom: BMJ Books.

Gucwa, K. J., & Cheng, H. H. (2014). Robosim for integrated computing and STEM education. In *American Society for Engineering Education Annual Conference and Exposition* (pp. 1–17).

Hamner, E., & Cross, J. (2013). Arts & Bots: Techniques for distributing a steam robotics program through k-12 classrooms. In *IEEE Integrated STEM Education Conference* (pp. 1–5). doi:10. 1109/ISECon.2013.6525207

He, S., Maldonado J, Uquillas A, & Cetoute T. (2014) Teaching k-12 students robotics programming in collaboration with the robotics club. In *IEEE Integrated STEM Education Conference* (pp. 1–6).

He, S., Zubarriain, J., & Kumia, N. (2015). Integrating robotics education in pre-college engineering program. In *IEEE Integrated STEM Education Conference* (pp. 183–188).

Jackson, J. (2013). An engineering mentor's take on: "first" robotics. *Tech Directions, 72,* 13–15.

Jeon, M., FakhrHosseini, M., Barnes, J., Duford, Z., Zhang, R., & Ryan, J., et al. (2016) Making live theatre with multiple robots as actors bringing robots to rural schools to promote steam education for underserved students. In *ACM/IEEE International Conference on Human-Robot Interaction (HRI)* (pp. 445–446). doi:10.1109/HRI.2016.7451798

Kaloti-Hallak, F., Armoni, M., & Ben-Ari, M. M. (2015a). The effectiveness of robotics competitions on students' learning of computer science. *Olympiads in Informatics,* 89–112.

Kaloti-Hallak, F., Armoni, M., & Ben-Ari M. M. (2015b). Students' attitudes and motivation during robotics activities. In *Workshop in Primary and Secondary Computing Education* (pp. 102–110). doi:10.1145/2818314.2818317

Karim, M., Lemaignan, S., & Mondada, F. (2015). A review: Can robots reshape k-12 STEM education? In *International Workshop on Advanced Robotics and its Social Impacts* (pp. 1–8).

Karp, T., & Maloney, P. (2013). Exciting young students in grades k-8 about STEM through an afterschool robotics challenge. *American Journal of Engineering Education, 4*(1), 39–54. doi:10.19030/ajee.v4i1.7857

Kazakoff, E. R., Sullivan, A., & Bers, M. U. (2013). The effect of a classroom-based intensive robotics and programming workshop on sequencing ability in early child-hood. *Early Childhood Education Journal, 41*(4), 245–255. doi:10.1007/s10643-012-0554-5

Khanlari A. (2013). Effects of educational robots on learning STEM and on students' attitude toward stem. In *Conference on Engineering Education* (pp. 62–66). doi:10.1109/ICEED.2013. 6908304

Kim, C., Kim, D., Yuan, J., Hill, R., Doshi, P., & Thai, C. (2015). Robotics to promote elementary education pre-service teachers' STEM engagement, learning, and teaching. *Computers & Education, 91,* 14–31. doi:10.1016/j.compedu.08.005

Kitchenham, B. A., & Charters, S. (2007). *Guidelines for performing systematic literature reviews in software engineering*. Technical Report, Evidence-based Software Engineering, United Kingdom

Larkins, D. B., Moore, J. C., Rubbo, L. J., & Covington, L. R. (2013). Application of the cognitive apprenticeship framework to a middle school robotics camp. In *ACM Technical Symposium on Computer Science Education* (pp. 89–94). doi:10.1145/2445196.2445226

Laut, J., Bartolini, T., & Porfiri, M. (2015). Bioinspiring an interest in STEM. *IEEE Transactions on Education, 58*(1), 48–55. doi:10.1109/TE.2014.2324533

Liu, A., Newsom, J., Schunn, C., & Shoop, R. (2013). Students learn programming faster through robotic simulation. *Tech Directions, 72*(8), 16–19.

Martin, T., Berland, M., Benton, T., & Smith, C. P. (2013). Learning programming with IPRO: The effects of a mobile, social programming environment. *Journal of Interactive Learning Research, 24*(3), 301–328.

McKay, M. M., Lowes, S., Tirhali, D., McGrath, E. W., Sayres, K., & Peterson, K. A. D. (2013). Transforming a middle and high school robotics curriculum. In *American Society for Engineering Education Annual Conference and Exposition* (pp. 1–21)

McKay, M. M., Lowes, S., Tirhali, D., & Camins, A. H. (2015) Student learning of STEM concepts using a challenge-based robotics curriculum. In *American Society for Engineering Education Annual Conference and Exposition* (pp. 1–25).

Merriam, S. B. (2009). *Qualitative research: A guide to design and implementation.* San Francisco, United States: Jossey-Bass.

Modekurty, S., Fong, J., & Cheng, H. H. (2014). C-STEM girls computing and robotics leadership camp. In *American Society for Engineering Education Annual Conference and Exposition* (pp. 1–14)

Montironi, M. A., Eliahu, D. S., & Cheng, H. H. (2015). A robotics-based 3d modeling curriculum for k-12 education. In *American Society for Engineering Education Annual Conference and Exposition* (pp. 1–14). doi:10.18260/p.23443

Mubin, O., Stevens, C., Shahid, S., & Al Mahmud, A. (2013). A review of the applicability of robots in education. *Journal of Technology for Education and Learning, 1,* 1–7. doi:10.2316/Journal.209.2013.1.209-0015

Nag, S., Katz, J. G., & Saenz-Otero, A. (2013). Collaborative gaming and competition for CS-STEM education using SPHERES zero robotics. *Acta Astronautica, 83,* 145–174. doi:10.1016/j.actaastro.2012.09.006

Nemiro, J., Larriva, C., & Jawaharlal, M. (2015). Developing creative behavior in elementary school students with robotics. *The Journal of Creative Behavior* 1–28. doi:10.1002/jocb.87

Nugent, G., Barker, B., Grandgenett, N., & Welch, G. (2016). Robotics camps, clubs, and competitions: Results from a US robotics project. *Robotics and Autonomous Systems, 75*(Part B) 686–691. doi:10.1016/j.robot.2015.07.011

Papert, S. (1980). *Mindstorms: Children, computers, and powerful ideas.* New York, United States: BasicBooks.

Petticrew, M., & Roberts, H. (2005). *Systematic reviews in the social sciences: A practical guide.* Oxford, United Kingdom: Blackwell Publishing.

Phamduy, P., Milne, C., Leou, M., & Porfiri, M. (2015). Interactive robotic fish: A tool for informal science learning and environmental awareness. *IEEE Robotics Automation Magazine, 22*(4), 90–95.

Piaget, J., & Inhelder, B. (1967). *The Child's conception of space.* New York, United States: Norton.

Pinzon, G. J., & Huerta, J. R. (2014). Introduction to STEM fields through robotics: A synergetic learning experience for students and their parents. In *American Society for Engineering Education Annual Conference and Exposition* (pp. 1–8).

Potkonjak, V., Gardner, M., Callagha, V., Mattila, P., Guetl, C., & Petrovi, V. (2016). Virtual laboratories for education in science, technology, and engineering: A review. *Computers & Education, 95,* 309–327. doi:10.1016/j.compedu.02.002

Prayaga, L., Prayaga, C., Wade, A., & Whiteside, A. (2013). The design and implementation of tele-robotics in education (TRE) to engage students in STEM disciplines: Including computer science and physics. *Journal of Computing Sciences in Colleges, 29*(2), 205–211.

Qidwai, U., Riley, R., & El-Sayed, S. (2013). Attracting students to the computing disciplines: A case study of a robotics contest. *Procedia—Social and Behavioral Sciences, 102,* 520–531. doi:10.1016/j.sbspro.2013.10.768

Rao, A. (2015). The application of LeJOS, LEGO Mindstorms robotics, in an LMS environment to teach children java programming and technology at an early age. In *IEEE Integrated STEM Education Conference* (pp. 121–122). doi:10.1109/ISECon.2015.7119902

Rubenstein, M., Cimino, B., Nagpal, R., & Werfel, J. (2015). Aerobot: An affordable one-robot-per-student system for early robotics education. In *IEEE International Conference on Robotics and Automation* (pp. 6107–6113).

Sahin, A., Ayar, M. C., & Adiguzel, T. (2014). STEM related after-school program activities and associated outcomes on student learning. *Educational Sciences: Theory and Practice, 14*(1), 309–322.

Sala, A. L., Sitaram, P., & Spendlove, T. (2014). Stimulating an interest in engineering through an "explore engineering and technology" summer camp for high school students. In *American Society for Engineering Education Conference and Exposition* (pp. 1–10).

Saleiro, M., Carmo, B., Rodrigues, J. M. F., & du Buf, J. M. H. (2013). A low-cost classroom-oriented educational robotics system. In G. Herrmann, M. J. Pearson, A. Lenz, P. Bremner, A. Spiers, & U. Leonards (Eds.), *International Conference on social robotics* (pp. 74–83). Springer International Publishing. doi:10.1007/978-3-319-02675-6

Sallee, J., & Peek, G. G. (2014). Fitting the framework: The STEM institute and the 4-H essential elements. *Journal of extension, 52*(2), 1–13.

Senaratne, H., Gunatilaka, P., Gunaratna, U., Vithana, Y., de Silva, C., & Fernando, P. (2014). SiFEB—A simple, interactive and extensible robot playmate for kids. In *International Conference on Artificial Intelligence with Applications in Engineering and Technology* (pp. 143–148).

Suescun-Florez, E., Iskander, M., Kapila, V., & Cain, R. (2013). Geotechnical engineering in us elementary schools. *European Journal of Engineering Education, 38*(3), 300–315. doi:10.1080/03043797.2013.800019

Sullivan, A., & Bers, M. U. (2013). Gender differences in kindergarteners' robotics and programming achievement. *International Journal of Technology and Design Education, 23*(3), 691–702. doi:10.1007/s10798-012-9210-z

Sullivan, A., & Bers, M. U. (2016). Robotics in the early childhood classroom: Learning outcomes from an 8-week robotics curriculum in pre-kindergarten through second grade. *International Journal of Technology and Design Education, 26*(1), 3–20. doi:10.1007/s10798-015-9304-5

Sullivan, F., & Heffernan, J. (2016). Robotic construction kits as computational manipulatives for learning in the STEM disciplines. *Journal of Research on Technology in Education, 48*(2), 105–128. doi:10.1080/15391523.2016.1146563

Talley, A. B., Crawford, R. H., & White, C. K. (2013). Curriculum exchange: Middle school students go beyond blackboards to solve the grand challenges. In *American Society for Engineering Education Annual Conference and Exposition* (pp. 1–10).

Tewolde, G., & Kwon, J. (2014). Robots and smartphones for attracting students to engineering education. In *Zone 1 Conference of the American Society for Engineering Education* (pp. 1–6). doi:10.1109/ASEE.2014.6820652

Toh, L., Causo, A., Tzuo, P., & Chen, I. (2016). A review on the use of robots in education and young children. *Educational Technology & Society, 19*, 148–163. doi:10.1080/15391523.2016.1146563

Tuluri, F. (2015). Using robotics educational module as an interactive STEM learning platform. In *IEEE Integrated STEM Education Conference* (pp. 16–20). doi:10.1109/ISECon.2015.7119916

Ucgul, M., & Cagiltay, K. (2014). Design and development issues for educational robotics training camps. *International Journal of Technology and Design Education, 24*(2), 203–222. doi:10.1007/s10798-013-9253-9

Yuen, T. T., Boecking, M., Stone, J., Tiger, E. P., Gomez, A., Guillen, A., et al. (2014). Group tasks, activities, dynamics, and interactions in collaborative robotics projects with elementary and middle school children. *Journal of STEM Education, 15*(1), 39–45.

Zelazo, P. D., Carter, A., Reznick, J. S., & Frye, D. (1997). Early development of executive function: A problem-solving framework. *Review of General Psychology, 1*(2), 198–226. doi:10.1007/s10643-012-0554-5

Chapter 6
Robotics Festival and Competitions Designed for STEM+C Education

ChanJin Chung, Christopher Cartwright and Joe DeRose

Abstract In order to promote and support STEM+C (Science, Technology, Engineering, and Mathematics plus Computing, Coding, or Computer Science) education, a student-centered robotics festival and competition called Robofest (www.robofest.net) was launched in 1999. Robofest's primary focus is the learning of STEM subjects together with computer science through autonomous robotics. When we make robots think, we will learn more because we have to think more. We believe programming team-built robots provide an effective environment to learn and exercise STEM disciplines in a truly integrated fashion. Furthermore, Robofest challenges are designed in such a way that dead reckoning is discouraged, which means students must program their robots with sensors to accomplish tasks in a dynamic and partially unknown environment. Through the challenges with unknown factors that require programming without adults' direct help, students learn, reinforce, and master STEM+C knowledge for twenty-first-century jobs. Robofest meets the needs of students based on their respective age, interest, learning style, and prior experience by offering diverse competitions such as Game, Exhibition, Vision Centric Challenge (VCC), Global Robotics Art Festival (GRAF), and Unknown Mission Challenge (UMC). As entry-level challenges for beginners, Robofest offers BottleSumo, RoboParade, and Carnival. After 17 years, there are currently over 2500 students participating in our programs annually in fifteen US states and fourteen other countries. Assessment and survey results have shown that the Robofest robotics experience has provided on opportunity for thousands of participants to learn more about STEM. Importantly, more students in the post-survey have indicated that they would consider a career involving STEM after Robofest exposure.

Keywords Robotics · Coding · Computing · Problem-based learning (PBL) · STEM+C

C. Chung (✉) · C. Cartwright · J. DeRose
Lawrence Technological University, Southfield, MI, USA
e-mail: cchung@ltu.edu

© Springer International Publishing AG 2017
M.S. Khine (ed.), *Robotics in STEM Education*,
DOI 10.1007/978-3-319-57786-9_6

6.1   Introduction

In 1967, Seymour Papert, Cynthia Solomon, Daniel Bobrow, and Wally Feurzeig crafted Logo, a revolutionary programming language, the first designed for use by children (Wikipedia 2016), at a time when mammoth computers occupied entire rooms. Papert's vision was that children should be programming the computer. At last, a half century later, programming languages are becoming non-foreign languages. Worldwide trends show that coding will become increasingly more important in early K-12 education. For example, in September 2014, the UK initiated the most ambitious attempt yet to get kids coding, with changes to their national curriculum that includes coding lessons for children as young as five (Dredge 2014). On December 10, 2014, US President Obama became the first president to write a line of code as part of the "Hour of Code"—an online event to promote Computer Science Education Week. The line of code in JavaScript he wrote was (Mechaber 2014):

$$moveForward(100);$$

Later, in January 2016 in his 2016 State of the Union Address, President Obama announced a "Computer Science (CS) for All" initiative to empower all American students from kindergarten through high school to learn computer science as a "new basic skill" and be equipped with the computational thinking skills they need to be creators not just consumers, in the digital economy, and to be active citizens in our technology-driven world (Smith 2016).

Robofest (www.robofest.net), launched in 1999, is an annual robotics festival with competitions designed to promote and support STEM+C (Science, Technology, Engineering, and Mathematics + Computing, Coding, or Computer Science) education (Chung and Sverdlik 2001; Chung and Anneberg 2003; Chung 2005, 2006, 2007, 2008, 2009, 2011, 2014 January, 2014 August, 2015, 2016; Chung and Cartwright 2010, 2011a, b, 2012, 2013; MacLennan 2010; Chung et al. 2012; Coscarelli 2015 March, 2015 September) with the following key strategies: (1) Autonomous robotics were chosen as a tool for STEM+C education because it provides a true hands-on learning environment and integrates all the components of STEM+C. (2) Robots were chosen, instead of simulation on a screen, as the target platforms to execute programs because Papert's research showed that children learned more efficiently if they could see a physical/tangible result for their computing efforts (Papert 1980). (3) Competition provides an active, collaborative, student-centered, and problem-based learning (PBL) environment. (4) Competition stimulates student motivation and performance. (5) Computer Science was the focus because the concept of "STEM" overlooked the importance of computer science, computing, programming, and coding at the K-12 level.

Robofest's mission is to inspire K-12+ students in STEM+C, develop teamwork, enhance creativity and problem-solving skills, and prepare them to excel in higher education and technological careers. Robofest challenges teams of students to design, build, and program autonomous robots to compete in various categories. Robofest has witnessed significant growth over the past 17 years. In the 2015–2016 season, 2575 students participated in Robofest. Cumulatively over 20,000 students have been a part of this program since its inception in the 1999–2000 school year at Lawrence Technological University (LTU) in Michigan. Robofest has drawn students from 15 States (Michigan, Ohio, New Hampshire, Texas, Florida, California, Washington, Missouri, Hawaii, Colorado, Indiana, Minnesota, Louisiana, Massachusetts, and New York) and 14 other countries (Canada, China, Colombia, Egypt, England, France, Ghana, Hong Kong, Hungary, India, Korea, Mexico, Singapore, and South Africa) and continues to expand at a rapid rate. What makes Robofest events unique? The following paragraphs summarize the unique characteristics of Robofest.

STEM+C Learning through autonomous robotics: When we make robots think, we learn more because we have to think more. Autonomous robotics is one of the best ways to learn all the STEM disciplines in a truly integrated fashion by focusing on computer science, computer programming, and computational thinking. Furthermore, Robofest challenges are designed in such a way that dead reckoning is discouraged. Students must program their robots with sensors to accomplish tasks in dynamic and partially unknown environments. Through specified challenges, students must apply the math and science skills learned in their classes to reinforce the learning.

Affordable: Our emphasis is on making Robofest affordable to all students, parents, and schools, and we accomplish this by charging a minimal team entry fee ($0–$50). Robofest does not require teams to hold major fund-raisers to underwrite the cost of participation. Robofest encourages recycling of all the logistic materials, such as robotic kits, parts, sensors, actuators, and playing field materials, which helps to control costs. It is easy and simple for organizations to host their own qualifier since the Robofest headquarters provide basic competition materials including trophies, medals, name badges, and certificates to them free of charge.

Flexible: Robofest allows students to use robotics kits and programming language of their choice. Any material can be used. The playing field materials are affordable, modular, and easy to transport and store, allowing student teams to practice anywhere at their convenience. Teams can be formed by any organization, school, homeschool, club, neighborhood, or civic group.

Students rule: While adult mentorship is encouraged, students are requested to design, construct, and program the robots themselves, and adult coaches/mentors are not allowed to assist during the actual competitions.

Everyone is recognized: All registered participants receive personalized medals and certificates if they do not drop out. Winners of the qualifying and championship rounds receive trophies. Top teams in the World Championship Sr. Division (grades 9–12) receive $2000 LTU renewable scholarships.

Free technical workshops are offered to teams on campus and/or online. All the workshop materials such as PowerPoint slides, example videos, sample programs, and recorded Webinars are available online for all teams. Some workshops are taught by high school seniors who have had extensive experience with Robofest.

Fig. 6.1 Diverse categories
of Robofest programs

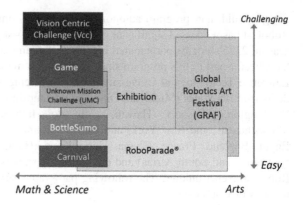

Various opportunities for every student: Robofest meets the needs of a variety of students based on their age, gender, learning style, and experience/ability levels. Figure 6.1 shows all eight Robofest program categories offered throughout the year and the different skill levels and interests.

Each category of Robofest is introduced in the following sections. Then, outcome and evaluation results will be summarized. A summary and conclusion will follow at the end of this chapter.

6.2 Game

Robofest Game, which is the most popular category in Robofest, is a timed mission defined with unknown factors and small unknown surprise challenges. Game is designed to adopt the problem-based learning (PBL) paradigm (Hmelo-Silver 2004). Learning is driven by challenging problems with no one "right" answer. Team members work as self-directed, active investigators, and problem-solvers in small collaborative groups. Teachers/Coaches adopt the role as facilitators of learning, guiding the learning process. Robofest Game especially puts math skills to the test. Teams need to apply math concepts with computational thinking skills to solve the missions. A new Game is introduced each year. The following subsections describe the recent year missions from 2012 until 2016.

6.2.1　2012 Game—R2R (Robots to the Rescue)

Game synopsis: *due to an earthquake, high-rise buildings (tissue boxes) in a city are on fire. An autonomous robot is being sent to rescue people (tennis balls covered with aluminum foil tape) on the top of the two buildings.* R2R playing field is shown in Fig. 6.2.

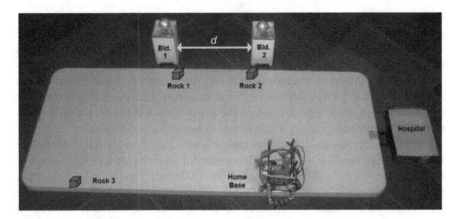

Fig. 6.2 2012 Game R2R (robots to the rescue) Jr. (grades 5–8) playing field

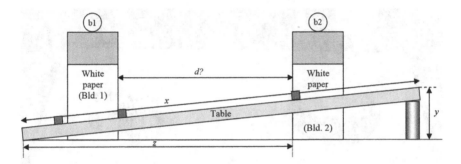

Fig. 6.3 2012 Game R2R (robots to the rescue) Sr. playing field

Missions of R2R Game are as follows: (1) Remove rock 3 off the table to clear the south edge road. (2) Move the rock in front of each building. (3) Rescue each individual (tennis ball) on each building. (4) Bring them into the hospital box. (5) Measure the distance between two buildings and report (display) the length in millimeters *at the Home Base*. Learning objectives of this Game are motion, manipulation, object detection, localization, logic, ratio, proportion, algebra, measuring, geometry, and navigation. High school teams were needed to use trigonometry especially, since the table was set up on an angle as shown in Fig. 6.3.

6.2.2   2013 Game—SRCC (Search, Rescue, Cleanup, and Collect Data)

An autonomous robot must search for and rescue people trapped in the black box in the tower of boxes, collect data, and clean up a contaminated area around the tower.

Detailed missions are to (1) remove (clean up) the white toxic boxes from the table, (2) bring the black box out of the contaminated area to home, (3) measure the size of the contaminated area in square millimeters and report/display the number, and (4) return to the Home Base.

For the Jr. Division, a black circle is used to represent the contaminated area as shown in Fig. 6.4. Two or three white boxes are used for the tower. The location of the black box is always on the top of the tower. The number of white boxes is unveiled 30 min before impounding robots to begin the competition.

For the Sr. Division, a right triangle shape is used instead of a circle. Two to four white boxes are used for the tower. The number of white boxes and the location of the black box in the tower are unveiled 30 min before impounding robots. It is known that the black box will not be at the bottom to make the challenge not to be too complicated. Teams are required to use geometry and/or trigonometry to measure the black shape. Sr. SRCC playing field is shown in Fig. 6.5.

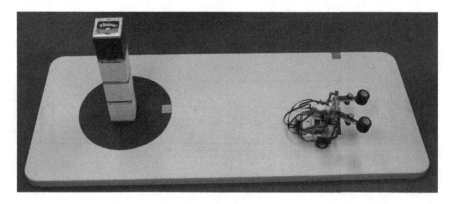

Fig. 6.4 2013 Game—SRCC (search, rescue, cleanup, and collect data) Jr. playing field

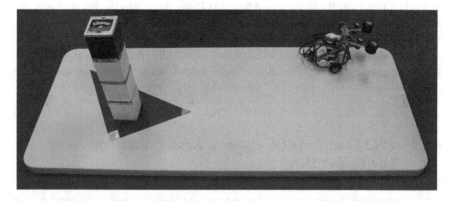

Fig. 6.5 2013 Game—SRCC (search, rescue, cleanup, and collect data) Sr. playing field

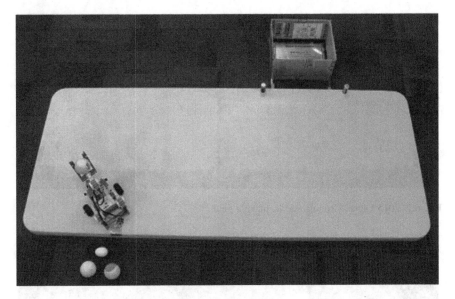

Fig. 6.6 2014 Game—avoid meltdown, Jr. playing field

6.2.3   2014 Game—Avoid Meltdown

A nuclear power plant is in trouble. An autonomous nuclear responder robot has detected the problem and has two minutes to deliver up to three water balls (tennis balls) and a special ball (hard-boiled egg) into the plant (box). The robot can carry only one ball at a time. Two concrete blocks (AA size batteries) near the plant need to be removed off the table. Also, the volume of the box (outer dimension) must be reported/displayed in cubic millimeters at the Home Base.

For the Jr. Division, the height and depth of the box are given. The box is aligned in parallel with the table. For the Sr. Division, only the depth of box is given and the box is not aligned in parallel with the table. Jr. Division playing field is shown in Fig. 6.6.

6.2.4   2015 Game—RoboBowl

The robot is to bowl, throw, shoot, or kick a tennis ball to knock down four pins (500 ml water bottles). If pins are knocked down, the highest point value will be awarded. If the ball just moves pins (but does not knock them down), partial points will be awarded. If the ball ends up in the pin side area, points are also given. In addition, the robot is required to report the height of the black rectangle shape on a letter size paper in millimeters. The location of pins 1 and 2 is unveiled. The robot must calculate the location of pins 3 and 4 based on the rectangle height measurement. Figure 6.7 shows an example of RoboBowl, Jr. playing field.

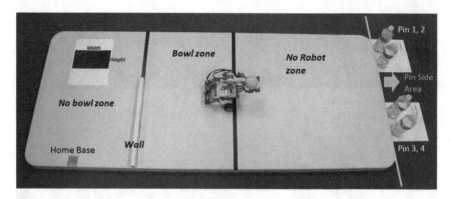

Fig. 6.7 2015 Game—RoboBowl, Jr. playing field

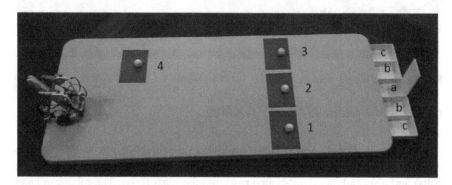

Fig. 6.8 2016 Game—RoboGolf, Jr. playing field

6.2.5   2016 Game—RoboGolf

There are four green areas with a golf ball. The robot must autonomously find each green area, locate a golf ball, stop, and putt the ball into a hole by using a specific piece of wood (wooden putter). The center hole, "a" in Fig. 6.8, has the highest point value. For the Jr. teams, the location of the four green areas will be unveiled at the competition site. However, for the Sr. teams, the exact location of the green area No. 4 will be completely unknown. The exact location of the golf ball on the green paper is unveiled to teams at the start of the competition.

During onsite workshops or Webinars, three methods to aim for the hole suggested were (Method 1) search for the flagpole by scanning using a sonar sensor, (Method 2) compute the location of the hole mathematically, and (Method 3) determine the location using trial and error—we can find the hole by rotating the robot different amounts in an attempt to find the correct orientation.

In the workshops and Webinars, the instructors reviewed the basic geometry needed to implement the (Method 2) mathematical approach. In order to aim the robot toward the center golf hole, the angle θ can be calculated using geometry as shown in

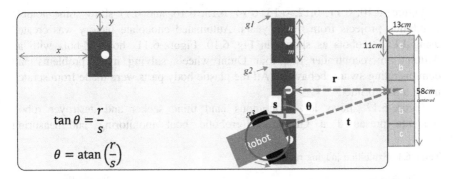

Fig. 6.9 2016 Game—RoboGolf, Jr.; to aim the robot toward the center golf hole

Fig. 6.9. Then actually the robot needs to spin, $90° - \theta°$. When the robot spins, the wheel path is a red circle centered between the wheels as shown in Fig. 6.9.

We explained an example to spin $90°$ when the robot's track width is 16.2 cm and the diameter of the robot's wheel is 5.5 cm. Students are instructed to calculate the circumference of the robot's path for a complete spin (the circumference of the red circle): Cp = PI * D = 3.14 * 16.2 cm = 50.87 cm. Then students are asked to calculate the circumference of the robot's wheel: Cw = PI * D = 3.14 * 5.5 cm = 17.27 cm. Since $90°$ is ¼ of a circle, the robot travels 50.87 cm/4 = 12.72 cm in order to spin $90°$. The final calculation is "How many wheel rotations are needed to travel 12.72 cm?" The number of rotations can be found by dividing the distance by the wheel circumference. Therefore, the answer is 12.72 cm/17.27 cm = 0.74 rotations.

6.3   Exhibition

Since the Game competition with fixed rules may limit students' creativity, Robofest offers a science fair-like stage for exhibitions to demonstrate creative robotics projects. The robotics Exhibition is a great way for students to show off their imagination and creativity. Each team has complete freedom to create any autonomous robotics projects such as robot pets, robots for scientific experiments, and practical robotics applications for any field. Computer-controlled robots with sensors may be of any size and can use any material as long as it is safe for team members as well as spectators. Hard-wired remote control is not allowed, but wireless host computer/robot control via software messages is allowed. Robot-to-robot communication is encouraged as well as human interaction with the robots. Suggested human interaction with robots includes: claps/knocks, flash light, color cards, and hand gestures. The application of math and science theories that are appropriate to the team members' age level is a strong plus for judging. Table 6.1 shows the summary of Exhibition judging rubric.

Figures 6.10, 6.11, 6.12, 6.13, 6.14, 6.15, 6.16, and 6.17 show some notable Exhibition projects from recent years. Automated chocolate factory was created using LEGO robots as shown in Fig. 6.10. Figure 6.11 shows Z-bots with an Arduino microcontroller and four Omni wheels solving maze problems and demonstrating swarm behaviors. All the plastic body parts were made from scratch using a 3-D printer.

Figure 6.12 shows an autonomous land mine seeker and destroyer robot. Students presented a GPS-guided robotic boat monitoring and measuring

Table 6.1 Exhibition judging rubric

Judging category	Weight (%)
Math and science concepts applied?	8
Students understand the math and science concepts applied?	8
Project idea	6
Originality	6
Robot demo performance	10
Project presentation	8
Project info (poster, brochure, Web site, video, etc.)	4
Teamwork	8
Robot mechanical design	7
New technologies, tools, parts, and materials	3
Project size	7
Practicality	7
Programming	8
Team independence	10

Fig. 6.10 Automated chocolate factory

Fig. 6.11 Arduino swarm robots

Fig. 6.12 Land mine seeker

ecological data (see Fig. 6.13). A student is playing a robotic violin (see Fig. 6.14) created with LEGO NXT controller and NXT sensors. Figure 6.15 shows intelligent four-way stop traffic control with self-driving cars built using Arduino controllers and ZigBee communication. Figure 6.16 shows a sensor-heavy smart robot arm built using Arduinos. A student is demonstrating a smart stick for the blind. It vibrates on the yellow line and it beeps when it detects obstacles (see Fig. 6.17).

Fig. 6.13 GPS-guided boat

Fig. 6.14 LEGO NXT violin

Fig. 6.15 Self-driving Arduino cars at four-way stop

Fig. 6.16 Sensor-heavy robot arm controlled by Arduino

Fig. 6.17 Stick for the blind

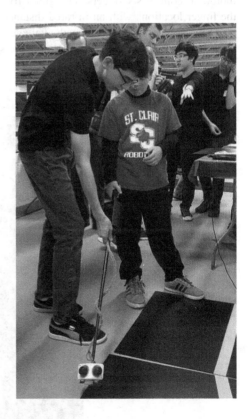

6.4   GRAF (Global Robotics Art Festival)

Robotics is all about STEM, and art is tightly coupled with all the STEM components. The idea of GRAF is to integrate the arts with robotics to provide effective and interdisciplinary STEM-learning environments where students will have an unforgettable "show and tell like" experience by creating robotics art projects using a universal language: art (Hamner and Cross 2013). GRAF implements the idea of "STEAM = STEM + A (Arts)" with two main categories: Performing Arts Division and Visual Arts Division. The Performing Arts Division includes dance/synchronized group dance, fashion show, music band, robot and human playing music together, robotic musical instruments played by humans (Chung et al. 2014 March; Chung 2014 March, 2014 July), and robot skit. Examples of Visual Arts Division are kinetic sculptures, kinetic canvas, and robotic painting. The following paragraphs introduce some GRAF projects.

Team Courageous 1 introduced a piano-playing robot and a robotic guitar instrument (see Fig. 6.18). The piano robot had a creative mechanism with three motors to control six fingers. The guitar robot made different tones while sensing the finger locations with an infrared distance sensor.

Team RoboCruisers introduced a believed-to-be the world's first robot playing a recorder (see Fig. 6.19). Three motors were synchronized to control two pumps to

Fig. 6.18 Piano robot and robotic guitar

Fig. 6.19 Recorder playing robot

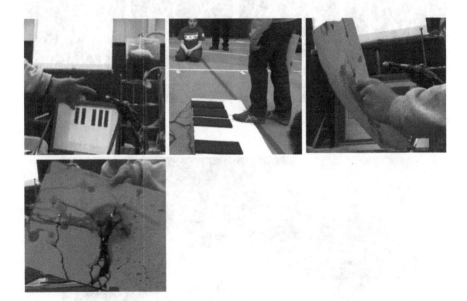

Fig. 6.20 Painting by mood

provide air powerful enough into the recorder windway to produce sound. When the robot plays "Mary had a Little Lamb" on a recorder, the other robot Big Bird dances to the music synchronized by Bluetooth signals.

Team Moodpainter from Mexico developed a robot system with a database that draws a shape by spraying different color paints according to a person's mood determined by the music played on a foot keyboard (see Fig. 6.20).

As shown in Fig. 6.21, team ALL2JESUS created a robot that displays kinetic art patterns by using 16 servomotors and distance sensors. The robot changes the pattern whenever it detects a new spectator or by using a built-in timer.

Team The Supernovas created a robot, Franc (see Fig. 6.22), which is designed to paint letters based on pre-entered coordinates. Franc painted the official logo for the global robotics art festival (GRAF) as shown in the figure.

Team Spruce Goose created an Arduino-based mechanical kinetic sculpture that abstractly depicts a cam shaft, in an artistic way, playing a LED light show activated by IR distance sensors (see Fig. 6.23).

Based on the assessment results published (Chung 2015), the GRAF has accomplished its goal to get students to pursue their interest in science, technology, engineering and math subjects, by using the power of universal human interest in arts. Student projects show hands-on application of STEM and computing science

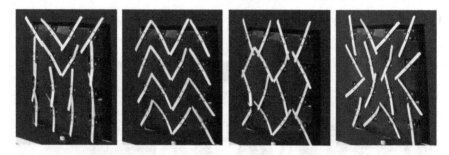

Fig. 6.21 Kinetic art patterns

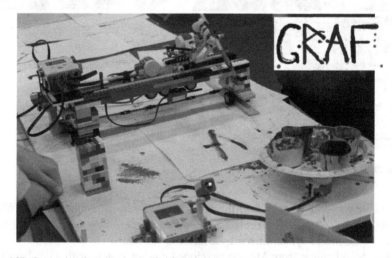

Fig. 6.22 Letter drawing robot and GRAF logo drawn by the robot

Fig. 6.23 Kinetic sculpture, Spruce Goose

skills to create robotics art projects that also require problem-solving skills. We also learned, as Gullatt argues, that arts are not only for (self) expression, but also for discovery (Gullatt 2008). Arts promote creativity while making everyone feel beauty and joy. Participation data show that the inaugural GRAF resulted in bringing more female and young students into STEM learning compared to Robofest competition populations (Chung 2014 January). The acronym for STEM + Arts is **STEAM**. Since robotics art is a typical example of STEAM, similarly, GRAF can be represented by **STREAM**, adding an R (Robotics) to STEAM.

6.5 Vision Centric Challenge (VCC)

Why is vision important to robots? Clearly, vision will enable a robot to become intelligent and autonomous in undertaking manipulation, navigation, and even social interactions. In this VCC launched in 2007, teams are required to build and program a robot with camera(s) to solve challenges that needs "seeing." This VCC is for advanced high school students as well as college students. Earlier challenges involved recognizing red/green light signals and green arrows, and following a lane while avoiding obstacles between two dashed lines (Crocker 2011) as shown in Figs. 6.24, 6.25, and 6.26.

Fig. 6.24 Red/green light
detection (Color figure online)

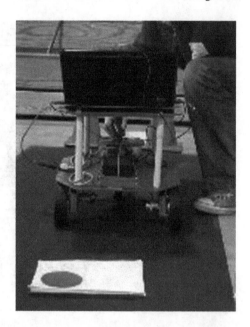

Fig. 6.25 Green arrow
detection (Color figure online)

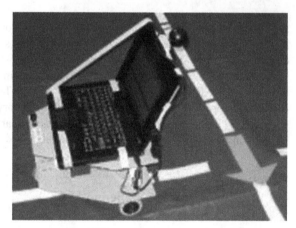

6.5.1 VCC 2014—Color Cup Navigation

To start the competition, a judge will show either Fig. 6.27 or 6.28 to the robot
camera. Then, the robot will be shown a digit on a letter-sized paper. After recognizing the digit, the robot should navigate through the path in such a way that the
blue cups are always on the left side if Fig. 6.27 is shown to the robot, for example.
If the number given was 2 for example as shown in Fig. 6.29, the robot needs to
return back home at the third yellow cross-line while maintaining the left blue color
rule (Fig. 6.30).

Fig. 6.26 Lane following while avoiding *orange cones* (Color figure online)

Fig. 6.27 *Blue* cup left (Color figure online)

Fig. 6.28 *Red* cup left (Color figure online)

Fig. 6.29 Pattern for number 2

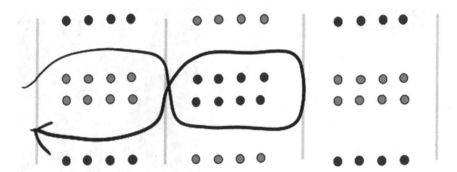

Fig. 6.30 Color cup navigation course example (Color figure online)

6.5.2 VCC 2015—MaxMin

The robot must follow a solid or dashed line to a series of waypoints (see Fig. 6.31). Each waypoint will be associated with a numeric digit. Robots must locate and traverse all waypoints recording their associated values during the challenge. Once all waypoints have been evaluated, the robot must return to the waypoint with the largest numerical value and spin 720°. The robot must then return to the waypoint with the smallest numerical value and stop. Waypoints for High School teams are represented by 9 × 12 in. sheets of colored construction paper. A table of the waypoint color to numeric value information is provided in Fig. 6.32. College waypoints are represented by white 8.5 × 11 in. sheets of printer paper. Each waypoint will have a printed orange shape. Paper and shape orientation will vary but remain consistent for all teams. All shapes with their associated numeric values are provided in Fig. 6.33.

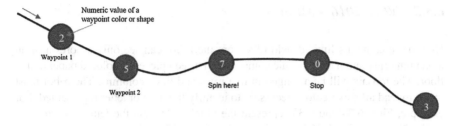

Fig. 6.31 Course with waypoints

Fig. 6.32 Waypoint numeric value chart for Sr. division

Paper Color	Numeric Value
Black	0
Blue	1
Brown	2
Green	3
Orange	4
Pink	5
Scarlet	6
Violet	7
White	8
Yellow	9

Fig. 6.33 Waypoint numeric value chart for college (Color figure online)

Shape		Numeric Value
Circle		0
Square		1
Triangle		2
Pentagon		3
Hexagon		4
Cross		5
Arrow		6
Moon		7
Heart		8
Star		9

6.5.3 VCC 2016—Mosaic

Due to the camera's limited field of vision, the robot can see only a portion of an alphanumeric pattern on a mosaic comprised of fifteen pieces of colored paper on the floor. The mosaic will be arranged in five rows and three columns. The robot must move to read all paper colors necessary to identify the digit or letter represented. For example, Figs. 6.34 and 6.35 represent the number "2" and the letter "A," respectively. Note that Fig. 6.36 is a varied pattern for A. The robot must report (display) the recognized digit or letter after spinning twice ($\sim 720°$) on the field.

Fig. 6.34 Pattern for "2"
(Color figure online)

Fig. 6.35 Pattern for "A"
(Color figure online)

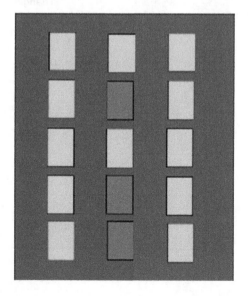

Fig. 6.36 Varied pattern for "A"

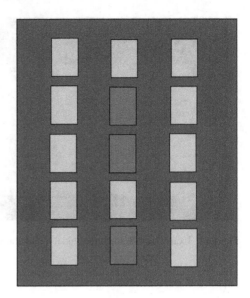

6.6   BottleSumo

The objective of the game is to either be the first robot to find and intentionally push a 2-L bottle (filled with 1 L of water) off the table or be the last robot remaining on the table. In either case, after the event (either the bottle was pushed off or the opponent are off the table), the robot must survive (remain) on the table for at least three (3) seconds. A robot is considered off the table when any of its parts are touching the floor, whether it was pushed off the table by the other robot or it fell off the table on its own. Each robot must be fully autonomous. No human control, signal, or remote computer control (tele-operation) is allowed. Through this entry-level BottleSumo, students can learn multiple STEM subjects such as physics, math, gears, logic, mechanical engineering, computer programming, and engineering design process by doing. The introduction of an additional target object, a 2-L bottle filled with a liter of water, makes the game even more challenging and minimizes the random chance of winning. Figure 6.37 shows Jr. BottleSumo starting configuration using a 6-ft table. The Senior Division field is made up of two tables.

6.7   Unknown Mission Challenge (UMC)

Mission tasks will be totally unknown until the day of competition. Contestants solve simple missions in a very short time frame (less than 3 h). The goal of this challenge is to provide an opportunity to develop problem-solving skills on the fly

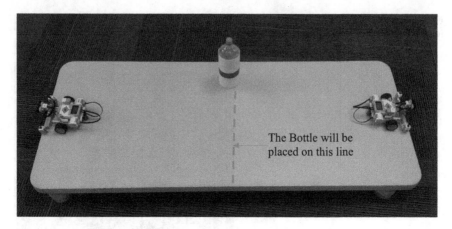

Fig. 6.37 Example of BottleSumo Initial configuration, Jr. division

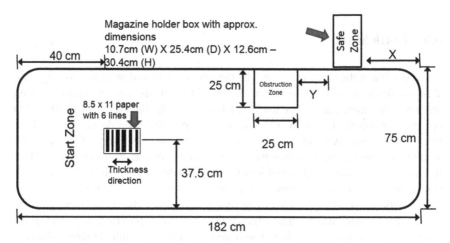

Fig. 6.38 UMC 2016 problem—read a bar code and deliver

without any help from adult coaches. Teams may use only approved robot kits. Pre-assembled robots cannot be used. Sensor or motor multiplexors are not allowed. The UMC 2016 Challenge problem was "read the bar code and deliver" (see Fig. 6.38). Your robot begins in the start zone loaded with a tennis ball and completes two tasks: (1) Report the total line thickness (sum of the individual six line thicknesses displayed on the robot screen) and (2) Deliver a tennis ball to the safe zone. The X value in Fig. 6.38 is 4 * (total bar code line thickness). We have witnessed and were truly amazed by the teamwork and solutions generated by the teams. The "parent-free" environment is enjoyable.

6.8   RoboParade

RoboParade, a parade of self-driving cars, is an entry-level program to engage, inspire, challenge, and prepare 4th–12th grade students for STEM+C learning and careers (Chung 2010; Chung and Cartwright 2014 January). RoboParade students design, build, program, and *decorate* autonomous robotic floats that follow a parade route, a black line. This fun event has featured miniature robotic floats with moving parts, sparkling lights, and all sorts of bells, drums, and musical instruments rolling down a parade route without remote controls. The students must program their robots to obey speed limits (minimum 7 cm/s and maximum 17 cm/s), display the current average speed, follow a black line, and avoid other robots using sensors. Through these tasks, students are learning, experiencing, and reinforcing STEM subjects such as proportion, arithmetic operation, arithmetic mean, linear function, unit conversion, ratio, circles, logic, data analysis, speed calculation, sensors, gears, motors, force, friction, center of gravity, and many others. To join the official parade, teams must pass a qualifying test that includes a written qualifying examination. The annual non-competitive RoboParade has been designed to complement the competition-oriented robotics events, targeting younger students and less-experienced teams by providing an entry-level robotics program into STEM, and freeing them of the stress of competition. RoboParade provides an interdisciplinary hands-on experiential approach where students may improve learning STEM subjects.

Robofest conducted a controlled experiment showing that students who participated in RoboParade 2012 and 2013 showed improvement on a pre–post STEM multiple-choice test compared to students who did not participate in RoboParade (Chung and Cartwright 2014). We believe the speed display requirement and the written qualifying test asking about speed calculation embedded into the RoboParade program contributed to the improvement of the RoboParade group. In addition, the artistic and non-competitive nature of the event seems to attract more female participants, with an over 28% female participation rate, well above the other STEM programs. RoboParade is expanding to other cities. Two cities in Michigan, one city in Florida, one city in Texas, and one city in Virginia have offered this fun and effective informal STEM and STEAM learning opportunities. Figure 6.39 shows the Halloween RoboParade 2015 at Macomb Community College in Warren, Michigan.

Fig. 6.39 RoboParade 2015 at Macomb Community College, Michigan

6.9 Carnival

Individual students will first learn how to program robots to follow a black line and to send Bluetooth messages to robots, then visit multiple stations (Programming/ Construction, Beginner/Advanced) to program the controller and play the games explained in the following paragraphs. Note that Carnival is not a team-based program.

Scorpion Balloon Blaster: If the controller is programmed correctly, students are asked to select a math or science trivia quiz card. If the answer is correct, then they will have the chance to control a LEGO scorpion robot via Bluetooth to pop a balloon. If the answer is not correct, the student may go back to the end of the line to try the trivia quiz again. Winners will be determined based on the completion time. See Fig. 6.40.

Goal Challenge (see Fig. 6.41): If the controller is programmed correctly, students are asked to select a math or science trivia quiz card. If the answer is correct, then they will have the chance to play with a LEGO soccer robot to kick tennis balls. If not correct, the child may go back to the end of the line to try the trivia quiz again. There will be 4 or more tennis balls with numbers on the field and some obstacles. The goal is to maximize the sum of balls successfully kicked into the goal. Some balls are easy to kick in, but the values are low. Only 2 min will be given for each player. Winners will be announced based on the scores earned.

Lifter Design and Race (see Fig. 6.42): If the controller is programmed correctly, students are asked to design a robot arm to lift a LEGO barbell. If a student brings back the barbell using the robot with the arm, the mission is accomplished. Winners will be determined based on the programming completion time and the game completion time.

Fig. 6.40 Scorpion Balloon Blaster

Fig. 6.41 Goal challenge

Fig. 6.42 Lifter design and race

Speed Calculation Challenge: After introducing the concept of speed/velocity calculation, a robot car will be started to follow a straight black line. When it stops at the red color tape at the end of the black line, the robot will display the time taken from the beginning. Students are asked to calculate the speed of the robot car in cm/s as well as in./s. Students will be given a tape measure. Winners will be determined based on the accuracy and use of correct mathematical formulas in the calculation. Additional recognition will be given if the student implements more complex line following algorithm with three states.

Block Math: Students are asked to calculate the number of LEGO blocks used to construct shapes/structures. A bonus problem involves calculating a gear ratio. Three minutes are given to complete the task. Winners will be determined based on (1) accuracy, (2) use of math formulas/calculations, and (3) time to calculate.

Slope Calculation: Each student is given a distance sensor connected to a LEGO computer. They must watch a presentation to understand how the sensor works. The goal is to calculate the slope of a secret board inside a black box using the distance sensor. Winners will be determined based on accuracy and the mathematical description used to solve the problem.

6.10 Outcomes of 2015–2016 Academic Year

In the 2015–2016 academic year, a total of 2575 students on 834 teams participated Robofest from eight countries (Canada, China, Colombia, Egypt, Ghana, Hong Kong, India, and South Korea) in addition to 13 States from the USA (California, Florida, Hawaii, Illinois, Massachusetts, Michigan, Minnesota, Missouri, New Jersey, North Carolina, Ohio, Texas, and Washington) (Chung 2016). Figure 6.43 shows the number of student participants since 2000. There was a surge in numbers this year due to the growth in the international sites especially in India. The cumulative number of registered students and teams in our Web database since 2000 has reached 20,569. Note that some of these students are duplicated from year to year.

The average Robofest team size in 2016 was 3.0 which is same as that of last year. We believe this small team size is good for effective learning, since each student has more opportunity to contribute to the team's objectives.

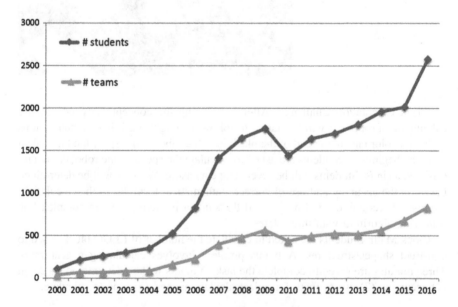

Fig. 6.43 Number of Robofest student participants and teams since 2000

Robofest offers a variety of categories in which to compete. Approximately 36% of teams participated in the RoboGolf Game in 2015–2016 year. The second most popular category was BottleSumo with 34%, then Exhibition with 15% of teams. Figure 6.44 shows percentages of teams by competition category. This does not include Robofest Camp (workshop plus mini competition) data.

Robofest competitions can be generalized into two categories: Games that use fixed rules including BottleSumo, VCC (Vision Centric Challenge), and UMC (Unknown Mission Challenge) and open-ended style with no fixed rules including Exhibition, RoboParade, and GRAF. Figure 6.45 shows the trend of number of teams between Games and Exhibition since 2005. We can see that the participation in the open-ended exhibition categories has been decreasing since 2013. We think students are more interested in fun game style challenges than creative science fair-like categories.

Robofest checks and records students' school grades, not ages. Approximately 40% of the students were from middle school, sixth through eighth grade. Approximately 27% were from high school, 17% were fifth graders, 15% were below fifth grade, and 1% were college students. Figure 6.46 shows the trend of each school grade (age) group since 2005. From 2015 to 2016, the percentage of fifth grade or younger increased from 22 to 32%. We believe this trend is positive since hands-on STEM+C learning opportunities should be offered earlier to talented young students.

Regarding gender, we experienced an increase of female student population in 2016; Approximately 71% were male and 29% were female students. Figure 6.47 shows the gender ratios of Robofest students. The average since 2005 is 75% male

Fig. 6.44 Percentages of teams per competition category in 2016

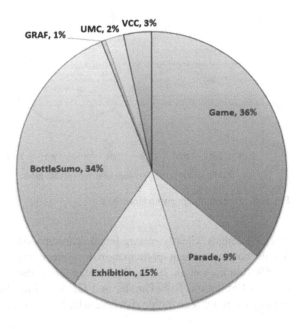

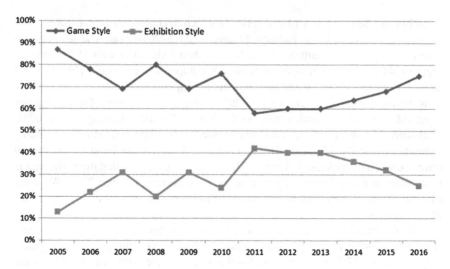

Fig. 6.45 Percentages of game and exhibition teams

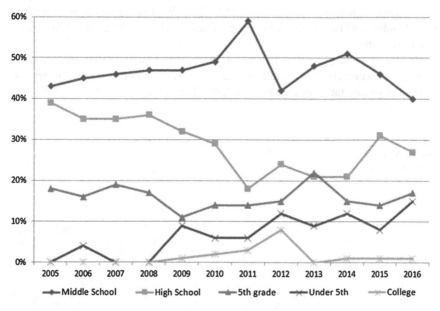

Fig. 6.46 Percent of age group since 2005

and 25% female. Clearly, young female participation in Robofest is far higher than female college student participation in college engineering programs, which is ~15% (Godfrey et al. 2010). We believe this is due to female friendly categories such as Exhibition, RoboParade, and GRAF. Note that again, the data is taken directly from our registration database which means it does not include the students

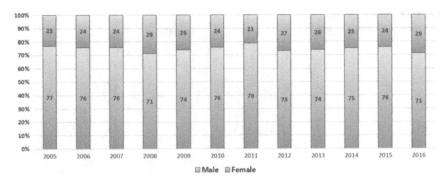

Fig. 6.47 Gender ratios of Robofest students

Fig. 6.48 Percentage of robotics kits used by teams in 2016

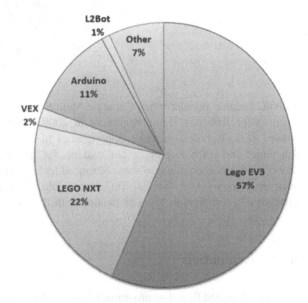

participating in Korea, Ghana, China, or Hong Kong as they were using their own registration system and did not provide us with their data.

Robofest is completely open and allows the use of any robotics platform, which is one of Robofest's unique features to promote creativity. Figure 6.48 shows the data on robotics kits used by the teams. Still the majority of the teams (79%) were using LEGO® products. The use of Arduino increased notably from 3% in 2015 to 11% in 2016. Other kits include Raspberry PI (Pandey et al. 2013).

Robofest remains focused on the student participants learning STEM+C through computer programming and testing. The programming languages used in Robofest 2016 are graphed in Fig. 6.49. Student teams continue to use advanced and varied forms of programming languages. Allowing students to use whatever programming language they prefer is one of the unique features of Robofest for STEM+C education. "Other C" in the figure includes Easy C, NXC, and Arduino C (Sketch).

Fig. 6.49 Percentage of programming languages used

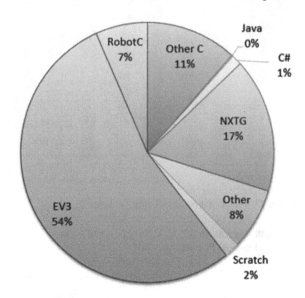

RobotC became popular when Carnegie Mellon Robotics Academy provided free licenses for Robofest teams beginning in 2009. All C-style languages together totaled 18%. Robofest provides opportunities to learn professional programming languages and helps to prepare our students for future professional career paths. Robofest students continue to show advanced technical skills and improvements in their problem-solving abilities. This is possible because of the many dedicated coaches and technical mentors associated with Robofest teams.

6.11 Evaluation

We have designed Robofest programs to collect data to evaluate the following three levels of impacts/outcomes (McLaughlin and Jordan 1999):

Short-term Impact: Acquired robotics knowledge and skills can be viewed as a short-term outcome/impact. Robofest is a perfect setting for students who want to be in the STEM+C career pathway as Robofest utilizes skills from all the STEM+C subjects. During the four months of the program, many students will learn for the first time in their lives how to write computer programs for real-time embedded control systems, which are their robots! Programming itself is not easy, but when their robots function correctly, it motivates students to work harder. In addition, they will learn many aspects of real-world engineering projects, which require problem specification, system design, implementation, and testing skills.

Intermediate-term Impact: Changes in students' behavior can be viewed as an intermediate-term impact/outcome of Robofest. Based on the knowledge and skills they learned and the real-world-like competition experience, their view of STEM+C-related classes will change. Students will have a reason to be more interested in learning about these subjects, and they will have changed their learning behavior in relating class subjects to real-world problems. As a result, they will be more attentive in their classes and have more confidence in their skills in these subject areas.

Long-term Impact: While they are participating in Robofest competitions, students who may have not yet decided on their career path could experience a life-changing revelation, which may result in a decision to study in STEM+C-related fields in college.

6.11.1 Short-Term Impact

To assess the short-term impact of Robofest roboticscompetitions, students were asked to take online assessments before and after the competition based on methodologies discussed in a previous study (Trudell and Chung 2009). In addition, the same assessment was taken with another group of students who did not participate in the competition as a control group. Each assessment consisted of 15 multiple-choice mathematics questions. Additionally, we collected information on the students' grade, gender, and whether or not they participated in Robofest, but no other identifying information. The pre- and post-assessments were implemented as a Google form.

Data (Chung and Cartwright 2011a, b) from Robofest 2011 involves a comparison of math scores among 4th–12th grade students who did and did not participate in either Robofest or other robotics competitions. The pre-assessment comparison comprised 164 students who participated in Robofest and 47 students who did not participate (the control group). The post-assessment comparison involved a subset of the students who took the pre-assessment: 51 Robofest students and 40 control students. The pre- and post-assessments were multiple-choice tests of 15 similar math questions. As shown in Fig. 6.50, Robofest students' mean

Fig. 6.50 Robofest 2011 math assessment results

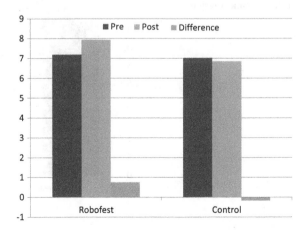

scores improved from 7.19 to 7.94 ($p < 0.10$), while the control group's scores actually decreased slightly.

In 2013 (Chung and Cartwright 2013), each assessment consisted of eight multiple-choice STEM questions (six math, one science, and one engineering). Additionally, we collected information on the students' grade, gender, and whether or not they participated in Robofest, but no other identifying information. The pre- and post-assessments were implemented as a Google document. Data from Robofest 2013 involves a comparison of math scores among 5th–12th grade students who did and did not participate in either Robofest or other robotics competitions. The pre-assessment comparison comprised 167 students who participated in Robofest and 104 students who did not participate (the control group). The post-assessment comparison involved a subset of the students who took the pre-assessment: 75 Robofest students and 102 control students. As shown in Fig. 6.51, Robofest students' mean STEM scores improved from 4.23 to 4.56 ($p = 0.19$) and STEM scores from students in the control group improved from 3.74 to 4.26 ($p < 0.10$). The higher participation rate from the control group (98% of the control group students took both the pre- and post-tests while only 45% of the Robofest students that took the pre-test also took the post-test) was a result that the control group took these assessments as part of their regular classroom, while the Robofest students took these assessments outside of the classroom.

2014 assessment (Chung and Cartwright 2014) consisted of seven multiple-choice STEM questions (five math, one science, and one engineering). Data from Robofest 2014 involves a comparison of math scores among 5th–12th grade students who did and did not participate in either Robofest or other robotics competitions. The pre-assessment comparison comprised 195 students who participated in Robofest and 61 students who did not participate (the control group). The post-assessment comparison involved a subset of the students who took the pre-assessment: 21 Robofest students and eight control students. Average school

Fig. 6.51 Robofest 2013 math assessment results

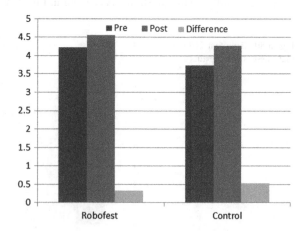

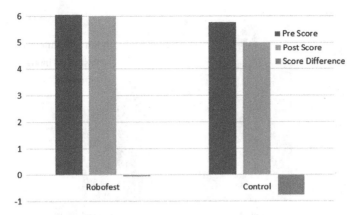

Fig. 6.52 2014 STEM score comparison between Robofest and control group

grade level of Robofest group was 7.52, while the control group average grade level was 10.8. Since the post-assessment was more difficult, both groups did not improve their scores. However, the Robofest group had higher scores, even if the average grade level is much younger and the control group performed worse on the assessment than Robofest group, as shown in Fig. 6.52.

These assessment studies in 2011, 2013, and 2014 suggest that participation in Robofest robotics competitions can help improve STEM scores (Chung et al. 2014; Chung 2015). We believe the use of explicit math components and unknown factors in Game designs as well as judging criteria requiring the use of mathematics and science components in Exhibition helped improve the STEM scores.

6.11.2   Intermediate-Term Impact

In 2014, we added a question to the pre- and post-assessments to check whether Robofest's experience would change students' interest in STEM subjects. Figure 6.53 strongly shows that Robofest experience impacted their preference toward STEM subjects as compared to a result of a control group. Robofest group's STEM interest increased from 77 to 96%; whereas the control group's STEM interest decreased from 63 to 50% (Chung and Cartwright 2014).

In 2015 and 2016 years, students were asked whether they would be interested in professional STEM and computer programing careers in the future. As shown in Fig. 6.54, 82.8% of Robofest 2015 students expressed their interest in STEM+C careers before participating in Robofest. After completing their Robofest participation for around four months, 88.7% of the students expressed their interest in STEM+C careers in the future (Chung 2015). The 2016 results showed similar impact, from 81.8 to 83.3% (Chung 2016).

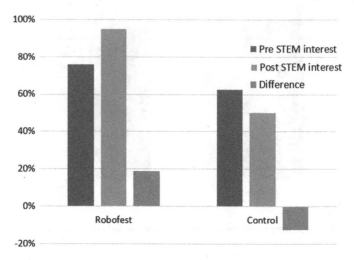

Fig. 6.53 2014 STEM interest comparison between Robofest and control group

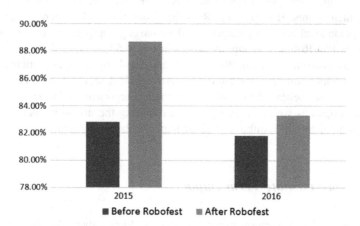

Fig. 6.54 STEM career interest

6.11.3 Long-Term Impact

To check how many students actually chose STEM career path, we contacted all the students who participated in the inaugural Robofest in 2000. Approximately 109 students participated in the first autonomous Robofest. Among them 12 Lawrence Tech college students entered the college level firefighting race, so they were omitted in our contact list. In 2011 after 12 years, using the phone number on the photograph release form (at that time we asked only phone numbers, no mailing address) we tried to contact 97 K-12 students, after verifying the last name with Internet sites such as whitepages.com. However, after more than a decade, we were able to contact only 14 of them (either the student or parent) successfully. Table 6.2

Table 6.2 College majors of inaugural Robofest students

Computer Science and Engineering	*2*
Mechanical Engineering	*2*
CAD (associate degree)	*1*
Biology	*1*
Civil Engineering	*1*
Business	2
Public policy	1
Film, broadcasting	1
English, marketing	1
Political Science	1
Did not go to college	1

The italic text denotes the STEM areas

Table 6.3 Characteristics of the eight Robofest competition categories

Competition Category	Rules	Unknown factors	Difficulty level	Art component	Age divisions
Game	Fixed	Some	Intermediate	No	Jr. and Sr.
Exhibition	Open-ended	N/A	All	Possibly yes	Jr. and Sr.
VCC	Fixed	Some	Advanced	No	Sr. and college
GRAF	Open-ended	N/A	All	Yes	Jr. and Sr.
RoboParade	Semiopen-ended	None	Beginner	Yes	fourth grade, Jr., and Sr.
BottleSumo	Fixed	Some	Beginner	No	Jr. and Sr.
UMC	Fixed but unknown	All	Intermediate	No	Jr. and Sr.
Carnival (individual)	Fixed	None	Beginner	No	fourth grade, Jr., and Sr.

shows the summary result. Approximately 50% of them (seven students) majored in STEM areas, 43% did not major in STEM, and 7% did not go to college (Chung and Cartwright 2011a, b). Table 6.2 also lists their majors.

6.12 Summary and Conclusion

Autonomous robotics can play an important role in STEM+C education because it naturally promotes hands-on learning and the integration of STEM subjects with computer programming and coding. Robofest has provided something for everyone, from novice to advanced, whether they like math and science subjects or not. Since Robofest challenges are relatively simple and small, iterative and incremental learning is practically possible within one season. Robofest introduces some new ways to use robotics such as VCC as an educational tool. Table 6.3 summarizes characteristics of the eight Robofest competition categories.

Robofest is a student-centered program and has generated enthusiasm among students. Programs are designed to adopt the problem-based learning (PBL) paradigm. Team members work as self-directed, active investigators, and problem-solvers in small collaborative groups up to five members. Teachers/Coaches work as facilitators of learning to guide the learning process.

Robofest has been operating successfully for seventeen years and has made broad impacts locally, nationally, and internationally. Every year, over 2500 students and over 800 teachers, coaches, parents, competition judges, and volunteers have been closely involved in Robofest programs. Robofest has been financially stable with support from team registration fees, sponsorships, and the donation of time by many volunteers. Robofest's eight innovative programs advanced STEM +C education by providing stimulating and diverse learning environments and by integrating STEM and Arts components through the use of computer science-oriented robotics. Robofest is an affordable and cost-effective program, and the ROI (return on investment) is relatively high (Chung 2015, p. 18).

The Robofest workshop curriculum focuses on mathematics and computational algorithms to solve the Robofest challenges. Assessment and survey results show that the Robofest robotics experience helps students learn more about Science, Technology, Engineering, or Math. Students expressed that they would consider a career involving Science, Technology, Engineering, or Math after participating in Robofest programs. Regarding long-term impact of Robofest programs, a preliminary survey shows that 50% of first-year participants in 1999–2000 majored in STEM areas when they attended college years after.

Author Note We would like to thank Robofest coordinators, Shannan Palonis and Katie Bis, for proofreading this article.

The following Robofest 2015–2016 year sponsors deserve recognition: Lawrence Technological University, Department of Math and Computer Science, DENSO, TOYOTA, LEGO Education, Michigan Council of Women in Technology Foundation, Nielsen, Robomatter, National Defense Industrial Association Michigan Chapter, IEEE Southeast Michigan Section, IEEE Region 4 PACE, RIIS, Realtime Technologies, Inc., mindsensors.com, Hanyang University, Aramark, CJ Chung, Howard Davis, ART/Design Group, Dennis Howie, and young+.

References

Chung, C. (2005). Robofest 2005 Annual Report. http://www.robofest.net/2005/05rpt2.pdf

Chung, C. (2006). Robofest 2006 Annual Report. http://www.robofest.net/2006/robofest06report. pdf

Chung, C. (2007). Robofest 2007 Annual Report. http://www.robofest.net/2007/robofest07report. pdf

Chung, C. (2008). Robofest 2008 Annual Report. http://www.robofest.net/2008/robofest08report. pdf

Chung, C. (2009). Robofest 2009 Annual Report. http://www.robofest.net/2009/robofest09report. pdf

Chung, C. (2010). RoboParade—An opportunity to develop your imagination while having fun. *Robot Magazine, November/December*, 48–49.

Chung, C. (2011). ROBOFEST 2010—Little robots perform big missions for STEM. *Robot Magazine, January/February*, 60–64.

Chung, C. (2014, January). Unveiling 2014 Robofest. *Robot Magazine, January/February*, 48–51.

Chung, C. (2014, July). The global robotics art festival (GRAF)—Robotics artistry through STEM. *Robot Magazine, July/August*, 60–62.

Chung, C. (2014, March). Integrated STEAM education through global robotics art festival (GRAF). In *4th IEEE Integrated STEM Education Conference (ISEC), Princeton University, NJ*, March 8, 2014.

Chung, C. (2014, August). Robofest 2013-2014 Annual Report. http://www.robofest.net/2014/robofest14report.pdf

Chung, C. (2015). Robofest 2014-2015 Annual Report. http://www.robofest.net/2015/robofest15report.pdf

Chung, C. (2016). Robofest 2015-2016 Annual Report. http://www.robofest.net/2016/robofest16report.pdf

Chung, C., & Anneberg, L. (2003). Robotics contests and computer science and engineering education. In *Proceedings of ASEE/IEEE Frontiers in Education (FIE) 2003 Conference, Boulder, CO*, November 5–8, pp. FIF-8–FIF-14.

Chung, C., & Cartwright, C. (2010). Robofest 2010 Annual Report. http://www.robofest.net/2010/robofest10report.pdf

Chung, C., & Cartwright, C. (2011a). Robofest 2011 Annual Report. http://www.robofest.net/2011/robofest11report.pdf

Chung, C., & Cartwright, C. (2011b). Evaluating the long-term impact of Robofest since 1999. LTU Technical Memo, ARISE-TM-2011-3. http://www.robofest.net/2011/ARISE-TM-2011-3.pdf

Chung, C., & Cartwright, C. (2012). Robofest 2011-2012 Annual Report. http://www.robofest.net/2012/robofest12report.pdf

Chung, C., & Cartwright, C. (2013). Robofest 2012-2013 Annual Report. http://www.robofest.net/2013/robofest13report.pdf

Chung, C., & Cartwright, C. (2014, January). RoboParade: A fun and effective way to promote STEM education. In *Proceedings of the 12th Hawaii International Conference on Education, Honolulu, Hawaii*, 5–9 January, 2014.

Chung, C., Cartwright, C., & Chung, C. (2014, March). Robot music camp: An experiment to promote S. T. E. M. and computer science. In *4th IEEE Integrated STEM Education Conference (ISEC), Princeton University, NJ*, March 8, 2014.

Chung, C., Cartwright, C., & Cole, M. (2014b). Assessing the impact of an autonomous robotics competition for STEM education. *Journal of STEM education, 15*(2), 24–34.

Chung, C., Marzougui, A., & Lahdhiri, T. (2012). Region 4 and the Southeastern Michigan section help foster STEM With Robofest PACE project. *IEEE USA in Action Magazine*, Spring 2012. http://www.nxtbook.com/nxtbooks/ieeeusa/ieeeusa_spring12/index.php?startid=29#/29

Chung, C., & Sverdlik, W. (2001). Robotics education in the K-12 environment, robots and education. *Stanford Spring Symposium (a part of the annual AAAI spring symposium series), Stanford University, Palo Alto, California*, March 26–28, 2001, pp. 37–45 (*).

Coscarelli, R. (2015, March). Robofest world championship. *Robot Magazine, March/April*, 20–23.

Coscarelli, R. (2015, September). Little robots—BIG missions. *Robot Magazine, September/October*, 28–31.

Crocker, N. E. (2011). Robofest 2011 vision centric challenge (VCC) update! *Robot Magazine, November/December*, 80–83.

Dredge S. (2014). Coding at school: A parent's guide to England's new computing curriculum. *The Guardian*. URL: https://www.theguardian.com/technology/2014/sep/04/coding-school-computing-children-programming

Godfrey, E., Aubrey, T., & King, R. (2010). Who leaves and who stays? Retention and attrition in engineering education. *Engineering Education, 5*, 26–40.

Gullatt, D. E. (2008). Enhancing student learning through arts integration: Implications for the profession. *The High School Journal*, 12–25.

Hamner, E., & Cross, J. (2013). Arts & bots: Techniques for distributing a STEAM robotics program through K-12 classrooms. In *Proceedings of the 2013 IEEE Integrated STEM Education Conference (ISEC)*, March, 2013.

Hmelo-Silver, C. E. (2004). Problem-based learning: What and how do students learn? *Educational Psychology Review, 16*(3), 235–266.

MacLennan, J. (2010). Robofest 2009—Motivating young minds to master the machine. *Robot Magazine, January/February*, 70–73.

McLaughlin, J. A., & Jordan, G. B. (1999). Logic models: A tool for telling your program's performance story. *Evaluation and Planning, 22*, 65–72.

Mechaber, E. (2014). President Obama is the first president to write a line of code. *White House Blog.* URL: https://www.whitehouse.gov/blog/2014/12/10/president-obama-first-president-write-line-code

Pandey, A., Wagle, R., & Brouillette, B. (2013). Raspberry PI robots & imaging processing. *Robot Magazine, November/December*, 56–59.

Papert, S. (1980). *Mindstorms: Children, computers, and powerful ideas*. New York: Basic Books Inc.

Smith, M. (2016). Computer science for all. *White House Blog.* URL: https://www.whitehouse.gov/blog/2016/01/30/computer-science-all

Trudell, E., & Chung, C. (2009). Development of methodologies to assess the impact of autonomous robotics competitions in science, technology, engineering, and math education. In *Proceedings of the International Technology, Education and Development Conference (INTED) in Valencia, Spain*, 9–11th of March. ISBN: 978-84-612-7578-6.

Wikipedia. (2016). Logo (Programming language). URL: https://en.wikipedia.org/wiki/Logo_%28programming_language%29

Chapter 7
Meeting Twenty-first Century Robotics and Automation Workforce Needs in the USA

Aleksandr Sergeyev

Abstract Recently, educators have worked to improve Science Technology Engineering and Mathematics (STEM) education at all levels, but challenges remain. Capitalizing on the appeal of robotics is one strategy proposed to increase STEM interest. The interdisciplinary nature of robots, which involve motors, sensors, and programs, makes robotics a useful STEM pedagogical tool. There is also a significant need for industrial certification programs in robotics. Robots are increasingly used across industry sectors to improve production throughputs while maintaining product quality. The benefits of robotics, however, depend on workers with up-to-date knowledge and skills to maintain and use existing robots, enhance future technologies, and educate users. It is critical that education efforts respond to the demand for robotics specialists by offering courses and professional certification in robotics and automation. The initiative presented here introduces a new approach for Industrial Robotics in Electrical Engineering Technology (EET) programs at Michigan Tech and Bay de Noc Community College. The curriculum and software developed by this collaboration of 2- and 4-year institutions match industry needs and provide a replicable model for programs in the USA. This project also addresses the need for certified robotic training centers (CRTCs) and provides curriculum and training opportunities for students from other institutions, industry representatives, and displaced workers.

Keywords Robotics · STEM education · Simulation software · Automation

7.1 Introduction

Many existing jobs will be automated in the next 20 years, and robotics will be a major driver for global job creation over the next five years. These trends are made clear in a study conducted by the market research firm, Metra Martech, "Positive

A. Sergeyev (✉)
Michigan Technological University, Houghton, MI, USA
e-mail: avsergue@mtu.edu

© Springer International Publishing AG 2017
M.S. Khine (ed.), *Robotics in STEM Education*,
DOI 10.1007/978-3-319-57786-9_7

Impact of Industrial Robots on Employment" (International Federation of Robotics: Metra Martech Study on Robotics 2016). Many repetitive, low-skilled jobs are already being supplanted by technology. However, a number of studies have found that in the aggregate, the robotics industry is creating more jobs than the number of jobs lost to robots. For example, the International Federation of Robotics (IFR) estimates that robotics directly created 4–6 million jobs through 2011 worldwide; with the total rising to 8–10 million if, indirect jobs are counted. The IFR projects that 1.9–3.5 million jobs related to robotics will be created in the next eight years (IFR International Federation of Robotics 2016). The rapid growth of robotics and automation, especially during the last few years, its current positive impact, and future projections for impact on the US economy are very promising. Even by conservative estimates (International Federation of Robotics: Metra Martech Study on Robotics 2016), the number of robots used in industry in the USA has almost doubled in recent years.

In the manufacturing sector, the recent growth was 41% in just three years—the number of robots per 10,000 workers employed in 2008 was 96 and reached 135 in 2011. The automotive sector in the USA relies heavily on robotics as well—China produces more cars than the USA, but the number of robots used in vehicle manufacture in China is estimated at 40,000 compared to 65,000 in the USA. From 2014 to 2016, robot installations are estimated to increase about 6% a year, resulting in an overall 3-year increase of 18% (International Federation of Robotics: Metra Martech Study on Robotics 2016). Likewise, industrial robot manufacturers are reporting 18–25% growth in orders and revenue year on year. While some jobs will be displaced due to the increased rollout of robots in the manufacturing sector, many will also be created as robot manufactures recruit to meet growing demand. Furthermore, jobs that were previously sent offshore are now being brought back to developed countries due to advances in robotics. For example, Apple now manufactures the Mac Pro in America and has spent approximately $10.5 billion in assembly robotics and machinery (Apple investing record $10.5 billion in supply chain robots & machinery 2016). In March 2012, Amazon has acquired Kiva Systems, a warehouse automation robot, and in 2013 deployed 1382 Kiva robots in three fulfillment centers. This initiative has not reduced the number of employees at Amazon; in fact, it added 20,000 full-time employees to its US fulfillment centers alone.

Such rapid growth of Robotic Automation in all sectors of industry will require an enormous number of technically sound specialists with the skills in industrial robotics and automation to maintain and monitor existing robots, enhance development of future technologies, and educate users on implementation and applications. It is critical, therefore, that educational institutions adequately respond to this high demand for robotics specialists by developing and offering appropriate courses geared toward professional certification in robotics and automation. In addition, certified robotic training centers (CRTCs) will be in high demand by industry representatives and displaced workers who need to retool their skills. The initiative described here will demonstrate and test an effective approach for teaching emerging topics of Industrial Robotics in Electrical Engineering Technology

(EET) programs at both the university and community college levels. The curriculum and software developed in this initiative between 2-year (Bay de Noc Community College) and 4-year (Michigan Tech) institutions matches current industry needs and provides a replicable model for the EET programs across the country. The project also addresses the need for CRTCs and provides curriculum and training opportunities for students from other institutions, industry representatives, and displaced workers. Resources developed via this project will be disseminated through a variety of means, including workshops, conferences, and publications.

7.2 Project Overview and Objectives

The overall goal of the project is to help meet the nation's forthcoming need for highly trained industrial robotics workers. Strategies include developing, testing, and disseminating an updated, model curriculum, laboratory resources, and simulation software package suitable for use in both 2- and 4-year EET programs. To complement this effort, outreach to K-12 students and teachers will work to enlarge the pipeline and diversity of students interested in careers in robotics. Programs will also be offered to students at other institutions and to workers in industry to broaden impact.

Specific project objectives include the following:

1. Provide Electrical Engineering Technology (EET) 2-year and 4-year students with current and relevant skills in Industrial Robotics by:

    a. Updating both the 2-year and 4-year EET curriculum to include skills in industrial robotics relevant to current industry needs.
    b. Enhancing the existing industrial robotics laboratory at Michigan Tech and establishing a similar laboratory at Bay de Noc Community College to demonstrate the value of state-of-the-art, hands-on training experiences and support the course changes.

2. Provide "stand-alone" programs to train and certify students from other institutions, industry representatives, and displaced workers.
3. Develop new "RobotRun" robotic simulation software and make it available at no cost for adaptation by the other institutions. This will allow current concepts related to industrial robotics to be taught even in locations without access to current robotics hardware.
4. Train faculty members at similar institutions to build expertise in industrial robotics using state-of-the-art FANUC Robots.
5. Develop a pipeline and encouragement for 2-year students (particularly underrepresented students, many of whom attend community colleges) to explore options in 4-year EET degree programs.

6. Conduct robotics-oriented seminars for K-12 teachers to expand their knowledge in engineering and science and increase the awareness of the role the field of robotics plays in STEM education.
7. Conduct robotic workshops for high school students to increase their interest in STEM fields, utilizing the appealing concepts of robotics and automation to attract participants.
8. Disseminate the new curriculum and software widely to significantly influence the future electrical engineering technology workforce by encouraging enhancements in other EET programs.
9. Survey of industry and K-12 partners about the importance of the proposed activities identifies a significant need for and interest in robotics curriculum development, industry worker training, and K-12 STEM outreach.

This project meets several objectives of educational curriculum reform in the USA by:

- Creating and widely disseminating improved and industry-relevant learning materials,
- Developing faculty expertise at both university and community college levels,
- Assessing the success of curriculum enhancements and the overall project in achieving goals,
- Providing career pathways for technicians from 2- to 4-year programs,
- Linking educators and programs in 2-year colleges with those in 4-year institutions and industry,
- Improving K-12 teachers' technological understanding and providing them with tools to engage students in real-world technological problems,
- Promoting STEM fields among K-12 students.

7.3 Educational Need

Robotics is a great tool to promote STEM fields. Educators have been making measurable progress toward improving STEM education from primary to tertiary levels of education, but challenges remain. Given the current shortage of student interest in STEM education, increased attention has been given to the appeal and attraction of robotics. The interdisciplinary construction of robots, which involves motors, sensors, and programming, makes it a useful pedagogical tool for all STEM areas. The novelty of robotics is instrumental in attracting and recruiting diverse STEM students. In the classroom, robotics can easily be used to introduce a variety of mandatory skills needed to pursue a variety of STEM career paths (Office of Science and Technology Policy, Domestic Policy Council 2006; Witeman and Witherspoon 2003; Mauch 2001). More specifically, a robotics platform advances students' understanding of both scientific and mathematical principles (Witeman and Witherspoon 2003), develops and enhances problem-solving techniques

(Rogers and Portsmore 2004; Beer, Chiel and Drushel 1999; Robinson 2005), and promotes cooperative learning (Nourbakhsh et al. 2005).

While robotics can be used as an interdisciplinary STEM learning tool, there is also a strong need for industrial certification programs in Robotics Automation. More and more robots are designed to perform tasks that people may not want to do, such as vacuuming, or are not able to do safely, such as dismantling bombs. They have changed the lives of Amyotrophic Lateral Sclerosis (ALS) patients by giving them the ability to speak after their vocal cords have failed, and have sparked our imagination in space exploration (not to mention our fascination with characters like R2D2). As many have put it, robots do our dirty, dangerous, and/or dull work. Millions of domestic/personal robots are already on the market worldwide, from lawn mowers to entertainment robots (Mauch 2001). As a result, popular interest in robots has increased significantly (Barker and Ansorge 2007; Barnes 2002; Papert 1980; Johnson 2012; Fernandez 2009; Downloads—World Robotics 2014 2016). Global competition, productivity demands, advances in technology, and affordability will force companies to increase the use of robots in the near future (Ciaraldi 2009a, b; Devine 2009a, b Schneider 2005). While the automotive industry was the first to use robotics, aerospace, machining, and medical industries now also rely on Robotic Automation (Morey 2007). More than ever, trained and certified specialists are needed to maintain and monitor existing robots and to develop more advanced robotic technologies (Ciaraldi 2009a, b; "Robots: More Capable, Still Flexible" 2005; Tolinski 2006; Devine 2009a, b).

As mentioned, robotics can be used as an interdisciplinary, project-based learning vehicle to teach STEM fundamentals (Ciaraldi 2009a, b; Alimisis 2005; Chang 2009; Liu 2009; Karatrantou and Tzimogiannis 2005; Eslami 2009; Ren 2009; Piaget and Piaget 1973; You 2009; Michalson 2009). Understanding the valuable role robotics education plays in helping students understand theoretical concepts through invention and creation, many universities include components of robotics research in curricular offerings (Ren 2009). It is widely recognized that robotics is a valuable learning tool that can enhance overall STEM comprehension and critical thinking (Ciaraldi 2009a, b; Piaget and Piaget 1973; You 2009). Because of these benefits and industry needs, new programs in Robotics Automation and applied mobile robots are popping up in the USA and abroad. Industrial help from Microsoft, FANUC Robotics America Inc., and MobileRobots Inc., is essential to the growth of these programs. The objectives behind robotic programs are clear: (1) In the short term, robotics education fosters problem-solving skills, communication skills, teamwork skills, independence, imagination, and creativity; and (2) in the long term, robotics education plays a key role in preparing a workforce to implement twenty-first century technologies.

Currently, few universities offer specific robotics degrees. For instance, Worcester Polytechnic Institute (WPI) has offered a Bachelor of Science in Robotics Engineering since 2007 (Guiding the Independent Learner in Web-Based Training 1999). Universities that have graduate degrees focused on robotics include Carnegie Mellon University, MIT, UPENN, UCLA, WPI, and the South Dakota School of Mines and Technology (SDSMT). Michigan State University has a well-established robotics and automation Laboratory, but it is utilized for graduate robotics courses and research. Very few universities across the USA offer a degree and/or certification specifically in Robotics Automation. In fact, Lake Superior State University (LSSU) is one of very few universities in Michigan that specializes in Robotics Automation; however, it does not have a program to certify industry representatives (Schneider 2005). With few focused industrial robotics programs, undergraduate industrial robotics training often occurs in electrical engineering technology programs, the focus of the proposed program. Training in Robotics Automation is especially important to Michigan's economy. A major decline in automotive manufacturing jobs has left many areas in Michigan with high rates of unemployment. Baraga County, located 15 miles south of Michigan Tech, has one of the nation's highest rates of unemployment. Yet, Michigan has an unmet need for workers in robotics jobs (Downloads—World Robotics 2014 2016). Filling these jobs, however, requires workers trained and certified in the following skill sets: designing, testing, maintaining, and inspecting robotic components; troubleshooting robot malfunctions; using microcomputers, oscilloscopes, hydraulic test equipment, microprocessors, electronics, and mechanics; and reading blueprints, electrical wiring diagrams, and pneumatic/hydraulic diagrams. Driven by industry needs, the new curriculum designed in this project will be adapted for both 2- and 4-year programs. The project aims to address the current US workforce need for properly prepared STEM professionals, train current industry representatives and displaced workers in Robotics Automation, educate K-12 teachers with the current art of industrial robotics, and promote STEM fields among K-12 students.

7.4   University and Community College Partnership

Michigan Technological University is a public university committed to providing a quality education in engineering, science, business, technology, communication, and forestry. Fall 2014 total enrollment was 7100 students, including 1442 (20.6%) graduate students. Over 65% of Michigan Tech students are enrolled in engineering and technology programs. Cumulatively, the School of Technology (SOT) and College of Engineering (COE) granted 700 undergraduate degrees in 2012–2013 with 63% of total first-major Michigan Tech undergraduate degrees. Twenty-two research centers and institutes support interdisciplinary research, partnerships with industry, and collaboration with community and informal education organizations. The SOT awards Bachelor's degrees in Computer Network and System Administration, Construction Management, Electrical and Computer Engineering Technology, Mechanical Engineering Technology, and Surveying Engineering—all degrees that require an understanding of robotics. Michigan Tech is rated highly for academics, career preparation, and quality of life in the Princeton Review's Best 379 Colleges 2015 Edition. Michigan Tech is ranked in the top tier of national universities according to the US News & World Report's "America's Best Colleges 2014." Employers, especially in the state of Michigan, have consistently relied on

Michigan Tech to deliver experiential educational opportunities. That is why Michigan Tech students average five interviews before they graduate, and why, despite record levels of unemployment in Michigan over the past three years, Michigan Tech has maintained on average a 93% job placement rate.

Bay de Noc Community College: While Bay College's racial and ethnic diversity is limited compared to other regions (Bay's student population is 86% White, 3% Native American, and 1.5% Hispanic), the socioeconomic reality of the local area is such that a large percentage of students enter college in a disadvantageous position. For instance, a full 58.4% of the Bay College's student population are first-generation students. Due in part to their lack of knowledge regarding the navigation of systems of higher education and in part to their typically lesser personal support systems, studies have shown that first-generation students have one of the lowest degree-completion rates of any demographic. A report published by the Higher Education Research Institute found that only 27.4% of first-generation students earn a degree after 4 years, compared to 42.1% of students whose parents have college experience ("Higher Education Research Institute," n.d.). While available data on the socioeconomic status of Bay's students is limited, the rate of Pell Grant eligibility, when used as a proxy, indicates that a large percentage of Bay College's students are low income. During the current fall semester, 62% of the Bay's student population were Pell eligible, representing another large population of students who enter college with a degree-completion disadvantage. This project will have a positive impact on the success of students by providing a local opportunity to receive for-credit education and training in a growing field. Students will have a clear educational pathway to either receive the necessary training or enter the workforce, or to articulate their educational experience to Michigan Tech for a more advanced degree. In either option, the clearly articulated pathway, the local availability of the program, and the many student support services offered by Bay College will help these students overcome obstacles and be successful in college and beyond.

7.5   Current Industry Partnership

Michigan Tech's EET department has an established collaborative relationship with FANUC Robotics America Inc., the leading company specializing in the development and production of innovative and intelligent robotic solutions. FANUC Robotics supports Michigan Tech's School of Technology, as well as other STEM programs. FANUC has provided significant educational discounts and made it possible to educational institutions to purchase resources that would otherwise not be feasible. Michigan Tech's FANUC Robot is the latest electric, servo-driven mini robot, housed in a self-contained, portable enclosure. It provides multiple benefits: industry-standard components that allow teaching principles of automation, compact and portable design, affordability, safe construction, and an integrated vision system that is commonly used in industry.

The FANUC Robotics Certified Education Robot Training (CERT) Program promotes understanding of FANUC Robotics' Automation solutions through the development and implementation of integrated classroom instruction and student projects. The CERT program is a new certification available to qualified universities. The program certifies instructors at educational institutions to train their students to program FANUC robots. Michigan Tech is a CERT-endorsed program. The EET department at Michigan Tech is a Certified Training and Education Site for FANUC Robotics Material Handling Program Software and iR-Vision 2D under the certification of Dr. Sergeyev. The collaboration between Michigan Tech and FANUC continues to bloom. In 2013 under Dr. Sergeyev leadership, Michigan Tech became a FANUC Authorized Certified Training Facility. Under this agreement, Michigan Tech becomes a regional training center specializing in industrial automation, eligible to train and certify students from other institutions, industry representatives, and displaced workers. Michigan Tech is one of only three existing FANUC Authorized Satellite Training Programs in the USA and the only one in the state of Michigan (F.A.S.T. 2014).

7.6   Current State of Robotics Automation at Michigan Tech and Bay de Noc Community College

At Michigan Tech, we are building a strong undergraduate Robotics Automation Program through the EET department. Dr. Sergeyev successfully launched the cross-disciplinary Robotics Automation-training program in 2009. The program continues to enroll: (1) Michigan Tech students from diverse STEM disciplines, (2) industry employees who seek retooling in Robotic Automation skills, and (3) students from other universities. Industrial collaboration between the EET department and FANUC Robotics is continuing to expand. The global presence of the automotive sector in Michigan, as well as the large number of companies specializing in industrial automation, generates an enormous demand for well-trained and certified in Robotic Automation specialists. Our ability to offer ongoing training and certification not only makes this project sustainable, but also provides us with an extensive opportunity for continuous evaluation of these educational initiatives. The current success of the program, CERT experts, and industry collaborations make Michigan Tech the ideal partner for Bay College. Via this joint collaboration, Bay College adapts Michigan Tech's existing curriculum in Robotics Automation, and implements and evaluates the robotics curriculum in a 2-year degree program. This partnership of 2- and 4-year institutions not only serves as an exemplary model of collaboration, but also allows for developing of a full spectrum of robotics curriculum adaptable at various levels of education. Bay College faculty collaborates with Dr. Sergeyev to adapt the existing curriculum for the 2-year program, develop new courses in Robotics Automation, and create state-of-the-art educational robotic training software. In addition, Bay College faculty participates

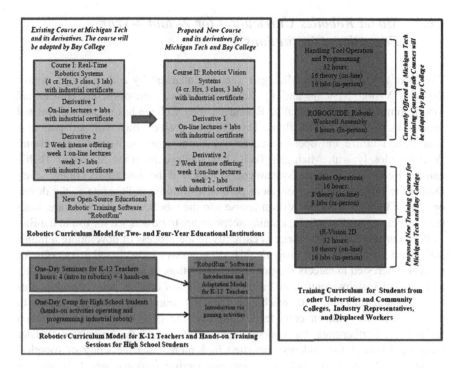

Fig. 7.1 Proposed Robotics Automation curriculum development

in the CERT program at FANUC. The goal is to become a FANUC-certified facility, which would allow Bay College to become a Certified Training and Education Site for FANUC Robotics Material Handling Program Software and iR-Vision 2D.

Robotics Automation Curriculum Development for EET Programs

This project has a significant impact on the curriculum at both institutions—Bay College and Michigan Tech. During this collaborative initiative, a broad spectrum of educational material is to be developed and made available between institutions for adaptation. Figure 7.1 depicts the proposed models in robotics curriculum development which will impact three different educational groups: (1) 2- and 4-year institutions; (2) students from other universities and community colleges, industry representatives, and displaced workers; and (3) K-12 teachers and high school students. The proposed curriculum changes are presented below.

7.6.1 Model Robotics Curriculum for 2- and 4-Year Educational Institutions

The robotics curriculum at Michigan Tech is enhanced by adding a new course Robotics Vision Systems and by developing new, state-of-the-art educational robotic simulation software called RobotRun. Bay College has no robotics-related courses in its current curriculum. Through this project, Bay will add two new courses and utilize the open-source robotic training software in the classroom. Working closely with Michigan Tech, Bay College modifies Michigan Tech's existing Real-Time Robotics System course for use at the community college level. During the project, faculty from both partner institutions collaborate in the development of the new Robotics Vision course and the robotics training software. Both the new course and the robotic training software are to be adapted for use in the Bay College curriculum. Bay College is currently in the process of developing a multi-skilled technician certificate program that will serve as a broad-based entry point for students interested in careers ranging from automation, to advanced manufacturing, to industrial systems control, and more. The robotics courses being developed as part of the partnership with Michigan Tech fits seamlessly into this degree to provide students with an entry-level understanding of robotics and automation technology. Furthermore, the courses included as part of the articulated transfer program designed for students interested in a more intensive study of robotics. These courses are also available to pre-engineering students, which is also an articulated transfer program to Michigan Tech. Students at Bay College benefit from these curriculum additions in numerous ways, including expanded exposure to advanced technologies, better career preparation, and increased options for transfer to Michigan Tech.

7.6.2 Building upon the Strong Foundation of Michigan Tech's Existing Course: Real-Time Robotics Systems

As shown in Fig. 7.1, the current curriculum at Michigan Tech in Robotics Automation includes one Real-Time Robotics Systems course (4 credit hour: 3 h of recitation and 3 h of weekly laboratory) covering all the theoretical and practical aspects of the knowledge required for technologists involved in the Robotics Automation industry. The course introduces the science and technology of mechanical manipulations and robotics systems control. A broad range of robotics topics is covered, including sensors, end effectors, and actuators. Essentially, this course is the building block for future coursework in the mechanics, control, and programming of robotic systems. Designed from a manufacturing perspective, the course addresses robots in an isolated manner while exploring the broad topic of industrial work cells that contain a robot. This includes robot automation and all related technology needed to integrate the robot with the work environment and with the enterprise database. The course also addresses the major aspects of design, fabrication, and robotic-enabling systems. Design aspects involve determining specifications for a robot, configurations, and what sensors and actuators should be used. Considerable attention is currently dedicated to safety procedures of operating robotics platforms. The currently used FANUC Roboguide software package allows students to learn the structure of the programming language commonly used in the field of robotics for off-line programming.

After receiving sufficient off-line programming training and passing the safety-related test, students implement their knowledge and perform laboratory experiments that involve hands-on programming and operation of a state-of-art FANUC industrial robot. This course offers the foundation of an educational platform for developing and implementing an effective curriculum model in Robotics Automation. Close collaboration with industry in the initial design of this course also helped to advance an industrial certification program that is endorsed by industry. Students who successfully complete the course are issued a FANUC industrial certificate in "Handling Tool Operation and Programming." To further promote the course development and make the course-offering model more flexible, two derivatives of the course have been developed. The first, a hybrid version of the course, has been successfully implemented several times since 2012. In this version of the course, the theory, quizzes, and examinations are delivered online, but students still have an opportunity for hands-on training during weekly 3-hour labortaories.

This model allows for more flexible scheduling of the class, which in turn helps students who work while attending school. The second-course derivative involves an intense 2-week structure with the same amount of theory and hands-on practices in a condensed time. The first week involves an introduction to the theoretical content, culminating in the midterm examination. In the second week, students are completely immersed in the hands-on activities of operating and programming FANUC industrial robots. The second week culminates with a 2-hour certification/final examination. The 2-week intense course model has proven to be very effective and has become very popular among students at Michigan Tech in a variety of disciplines. Since 2009, Dr. Sergeyev, a FANUC-certified instructor, has trained and awarded industrial certificates to more than a hundred students. Building upon the foundational course in place at Michigan Tech, Bay College adapted and implemented all the derivatives of the Real-Time Robotics System course in order to provide Bay students with flexible course offerings and certification options.

7.6.3 Robotic iR-Vision 2D Course

Nearly any robot currently used in industry is equipped with a vision system. Vision systems are being used increasingly with robot automation to perform common and sometimes complex industrial tasks, such as part identification, part location, part orientation, part inspection, and tracking. The vision system provides the robot "eyes" needed to perform complex manufacturing tasks. The Robotic iR-Vision 2D course is designed as a four-credit hour course (3 h of recitation and 3 h of weekly laboratory). The course introduces topics on the following: (1) safety, including laser safety; (2) basics of optics and image processing; (3) setting up lightning conditions required for the successful vision error proofing and camera calibration; (4) teaching tool, application, and calibration frames; (5) performing 2D calibration and 2D single and multi-view robotic processes; (6) performing 3D calibration and 3D single-view robotic vision processes.

Hands-on training is an integral part of any course developed in the School of Technology at Michigan Tech, and this course is no exception. It includes 12 laboratory exercises, totaling 36 h, with the goal of providing students the opportunity to configure and execute real-life, industry comparable, and robotic vision scenarios. The course is similar to the existing Real-Time Robotics Systems' rigorous assessment strategy and culminates in a 2-hour certification examination. Students successfully passing the examination receive a certificate in iR-Vision 2D. In addition to the traditional offering, two derivatives (a hybrid and 2-week intense version) of the Robotics iR-Vision 2D course have been developed at both institutions—Bay College and Michigan Tech.

The industrial automation laboratory at Michigan tech has four FANUC training carts each comprising of a FANUC LR Mate 200iC robot, R-30iA Mate Controller, Sony XC-56 camera, air compressor, and a computer. These robots have an option for interchangeable end effect or such as suction cups and 2-finger parallel grippers, which are used in developing a variety of applications. The iR-Vision 2D course offered to students at Michigan Tech consists of 12 laboratory exercises that help them gain hands-on training and experience with the FANUC iR-Vision 2D system. A scaled down version of the same set of laboratory exercises is used for the certification program for industry representatives.

Following are the topics for the laboratory exercises:

Camera and lighting concepts,
Camera setup
Frames,
2D calibration,
Error proofing,
2D single-view process,
2D single-view process: pill sorting,
2D single-view process: chips sorting and palletizing,
2D single-view process: battery picking, orienting and placing.

The first few exercises begin with introducing the students to all the hardware and software components of the setup consisting of the camera, lights, robot controller, and iR-Vision software. Significant attention is given explaining the wiring and communication between the camera, robot controller, and the computer. The above also includes the type of camera and connection ports and explains the procedure to setup the camera. Setting up the camera allows the user to select the exposure time and type of mounting of the camera and attain different parameters such as image size and aspect ratio. TCP/IP is the protocol used between the controller and the computer to communicate through the software, and the hands-on activity is designed to provide a stepwise procedure to achieve successful communication. After the communication is established, the user can access the software and learn the functionality of all the options on the graphical user interface of the software. Objects that will be used to be taught and further recognized by the robotic vision system are placed in the camera's field of view. The camera view is obtained via the robotic vision software and respectively displayed on the computer screen. The clarity of images can be improved by varying the contrast and the exposure time of the camera. These exercises help students to understand how the camera perceives images using pixels under different lighting conditions and apply this knowledge in future.

The next stage is to teach students the coordinate systems referred to as frames, related to the robot and camera's environment. The three frames that affect the motion of the robot are world, user, and tool frames. Using the teach pendant, a handheld device used to program and control the motion of the robot, these frames are taught to the robot and used in the procedure for camera calibration. Camera calibration helps in locating the position of the camera with respect to the robot world frame by implementing a calibration grid, a predetermined pattern of black circles drawn in a grid format, and helps determine different parameters such as focal distance and location. After teaching the basics of setting up and calibrating the vision system, the process of error proofing is introduced in the next session. "Error Proofing in automation relates to the ability of a system to either prevent an error in a process or detect it before further operations are performed (FANUC Robotics System R-30iA Controller iR-Vision with Error Proofing Student (Manual), n.d.)." It is widely used in the industry for various applications and can be performed on manufactured parts in a process, or can be used to monitor critical components of a process.

Error proofing technique identifies the presence/absence or orientation of parts, and critical areas on a part and is an economical way to perform quality checks (FANUC Robotics System R-30iA Controller iR-Vision with Error Proofing Student (Manual), n.d.). "The Error Proofing Process requires no calibration and does not return any part offset that can be used to modify robot movement. It does, however, return a pass or fail dependent on criteria set by the user (FANUC Robotics System R-30iA Controller iR-Vision with Error Proofing Student (Manual), n.d.)." The process involves stepwise approach to teach different objects' location, orientation, and size to the vision system. It uses the geometric pattern-matching tool to teach the pattern of the object, and this tool includes features such as masking and emphasis area, which help in identifying unimportant areas of the image or emphasize on important ones. Image recognition accuracy is expressed by a score threshold, and the target object is successfully found if its score is equal to or higher than this threshold value. Based on the PASS and FAIL results of the error proofing process, the user can use this data to program the robot.

Upon going through initial laboratory exercises described above, students obtain hands-on training on the basics of robotic vision and become well accustomed to the vision system process. The next few sessions involve the use of 2D single-view process with the camera in a fixed mounted position, and different practical applications using the vision system are programmed on the robot. One of the exercises is a pill sorting application in which few pills of two different colors (red and white) are placed on a black background (Fig. 7.2) and two empty bottles are placed on the side for collecting them. The main objective of this exercise is to recognize the differently colored pills using the vision system, and pick and place them into their respective bottles. The robot equipped with vacuum cup end effect or uses suction for the pick and place process. This exercise trains the students in differentiating between objects of the same size but different colors and improves their programming skills of using iR-Vision. Palletizing is a process of stacking products on to a pallet with a defined pattern of forming the stack. It is a widely used application in the industry, and the next laboratory exercise integrates the vision system techniques with the palletizing option installed on the controller of the robot. Round chips with different numbers printed on them are placed randomly (Fig. 7.3) on the base frame, and each chip is taught as a different object pattern to the vision system. First program is written on the teach pendant to locate the position of these chips and pick them up using suction cups of the robotic end effector.

A second program is written using a preinstalled option on the teach pendant called Palletizing EX. This option teaches the approach, picks up and places points to create a vertical stack of chip in a desired matrix format at predefined locations and places chips at corresponding positions. The students use their programming skills to integrate the above-mentioned programs and execute the desired objectives.

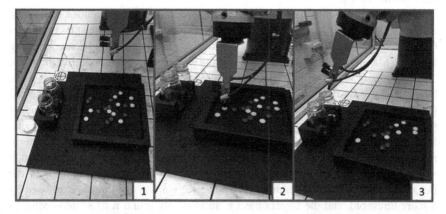

Fig. 7.2 (1) Randomly placed *red and white pills*, (2) robot pickup position, (3) *white pills* drop position

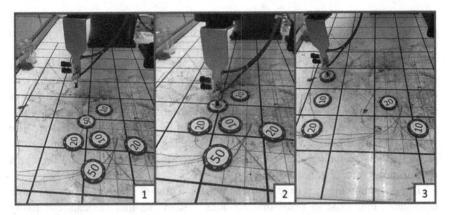

Fig. 7.3 (1) Randomly placed numbered chips, (2) robot pickup position, (3) robot orients and places chips at corresponding positions

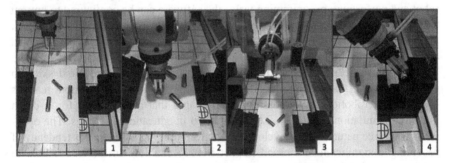

Fig. 7.4 (1) Randomly placed batteries, (2) pick position, (3) orientation check position, (4) drop position

The objective of the next laboratory is to recognize the position and orientation of a set of randomly placed batteries, pick them up one at a time, show the positive terminal of the battery to the camera to check for orientation of the battery, and drop them in a given slot with the positive terminals of all the batteries on one side. The setup is provided with the camera placed above the area and all the stepwise functions required to be complete the task are shown in Fig. 7.4.

7.6.4 New Open-Source Robotic Training Software "RobotRun"

A team of undergraduate students led by faculty members of Michigan Tech is developing an engaging, free, and open-source robotic training software aimed at helping students learn the basics of programming robotic arms. The software acts as a simulator where a user can write a program and then view how that program performs when run on a virtual 3D robotic arm displayed on the screen.

Although robotics play an essential role at a variety of manufacturing facilities, there is currently no accessible and free software that can give students the opportunity to learn about robotics and its applications. The developed software is intended to be used alongside the other training materials developed as a part of this project, but also be available online for anybody to download and use. The RobotRun software shows a 3D, animated rendering of a robotic arm that can be controlled via an intuitive programming language that is similar to the programs used to program real robotic arms.

The programming language of the simulator software provides all of the basic commands that exist in real-world robotics systems so that students can easily transfer the knowledge gained from the developed software to real-world robotic arms. The software allows the user to control where the end effector should jog, at what speed and type of motion termination, how many times it should repeat the movement, and other common robotics controls. Besides the option of jogging the robot and performing programming tasks, the software can be configured to present users with different scenarios that mimic real-world industrial scenarios such as pick and place, palletizing, welding, and painting. The program also allows users to load and save their programs so that they can turn them into an instructor for grading. The new software provides all options necessary to teach the required skills in robotics handling tool operation and programming. It is intuitive, without features of expensive robotic simulation software packages that are designed for in-depth industrial simulations and are not typically used in educational settings. The open-source and free nature of the developed RobotRun training software has tendency of providing a significant and broad impact by: (1) enabling institutions unable to obtain expensive industrial robots to adapt and teach the developed robotics courses; (2) providing K-12 teachers with the opportunity to promote STEM education to students by introducing the appealing concept of robotics via an interactive training environment, at no cost to K-12 institutions; (3) providing displaced workers wanting to improve their robotics skills with an intuitive, interactive and complete tool to succeed.

7.7 Curriculum for Students from Any Institution, Industry, and Displaced Workers

While robots play a role in all STEM fields, robots are key components of most manufacturing industries—from health to automotive sectors. Robotic Automation has been embraced as a way to stay globally competitive and to reduce the reliance on manual labor to perform redundant tasks. If the USA does not want to outsource, we need to automate. To provide support for the industry, educational institutions need to: (1) develop a training curriculum with industrial certification available to students from institutions where a robotics curriculum is not available; this will make those students more valuable in the job market; (2) provide effective, certified training to industry representatives who need to retool their skills to match rapidly developing technologies, especially in the field of Robotics Automation; (3) provide displaced workers with the opportunity to enhance, or acquire new, skills in robotics and enter the in-demand robotics job market. Michigan Tech's existing industrial certification program is enhanced by offering two additional FANUC certificates.

7.7.1 Certification 1: Handling Tool Operation and Programming (32-Hour Course)

The course is designed to be both practical and progressive. The content offers easily applied guidance to personnel involved in manufacturing with current robotic systems on site, or who may be asked to engage in implementing robotic systems in the near future. The course includes a discussion of scholarly and practical robotic topics ranging from kinematics and programming to practical application areas and economic concerns. Topics include the development of industrial robotics; an overview of the mechanical design, control, programming, and intelligence; organizational and economic aspects; robotics in operation and various applications. Hands-on experience is an essential part of this course and will occupy 70% of course time. The laboratory exercises are devoted to practical aspects of programming FANUC Robotics robots. This 32-hour course is designed to be offered partially online. The first 16 h are devoted to theoretical content delivered online. The remaining 16 h provide extensive hands-on experience working in the laboratory at Michigan Tech manipulating and programming FANUC industrial robots.

The course culminates in a certification examination in which the participants will have to demonstrate an understanding of theoretical background as well as the ability to program the robot for a task given by the instructor. Participants successfully passing the examination will receive a certificate issued by a FANUC-certified instructor. Due to the nature of the course, it can be offered on demand and conducted during weekends, students' breaks or in the summer. This flexibility has proven successful in attracting students from other institutions, industry representatives, and displaced workers. Feedback from past participants of the Michigan Tech training sessions showed very positive results. Students indicated that the partially online delivery approach not only saves travel time and money, but also allows participants to be more focused on the hands-on part of the course, thereby providing a more effective learning environment.

7.7.2 Certification 2: Robo Guide—Robotic Work Cell Assembly (8-Hour Course)

FANUC Robo guide software is widely used in industry; therefore, there is a great need to train workers in this software. The developed an 8-hour training course provides participants with a foundation for understanding all software features. By the end of the course, students assemble a fully functional virtual robotic work cell that includes the robot, end effector, several fixtures, and industrial parts that the robot can manipulate. Students program the robot to execute "pick-and-place" operation, run simulation in step-by-step and production modes, and compile a file that can be further transmitted to the physical FANUC robot for real-time production. This one-day training can be offered on demand and in conjunction with the other existing and under development certification courses.

7.7.3 Certification 3: Robot Operations (16-Hour)

There is a great demand in the industrial sector for robot operators that do not necessarily need to have very in-depth programming and theoretical skills. This course is intended for the person who operates or may be required to perform basic maintenance on FANUC robots via the standard application software package. It teach students how to safely power up, power down, jog the robot to predefined positions, and set up different frames of operation. In addition, it will cover tasks and procedures needed to recover from common program and robot faults, and teach basic programming skills. The course does not address the setup and operation of specific software features and options nor will it teach in-depth programming skills. These are covered in the 32-hour Handling Tool Operation and Programming course.

7.7.4 Certification 4: iR-Vision 2D (32-Hour Course)

Any robot currently used in industry is equipped with a vision system. Vision systems are being used increasingly with Robot Automation to perform common and sometimes complex industrial tasks, such as part identification, part location, part orientation, part inspection, and tracking. In other words, the vision system is the robot's "eyes" needed to perform complex manufacturing tasks. This developed course teaches students how to set up, calibrate, teach, test, and modify iR-Vision applications using FANUC robots. The course includes detailed discussion of hardware and software setup, establishing the communication link between the robot and teaching computer, teaching single- and multi-view processes, and programming. Safety procedures will be integrated into all exercises. As an integral part of this course, a series of laboratory exercises will be developed to provide hands-on training to reinforce the theory the student has learned. This 32-hour course is designed with a structure similar to the Handling Tool Operation and Programming course: 16 h of online and 16 h of hands-on training. The course culminates in a certification examination in which the participants demonstrate an understanding of the theoretical background, as well as the ability to successfully set up, calibrate, program, and utilize the FANUC robot equipped vision system. Participants passing the examination receive a certificate in iR-Vision 2D issued by a FANUC-certified instructor. Similar to the other certification courses, it can be offered on demand and conducted during weekends, students' breaks or in the summer.

7.8 Task 3: Model Robotics Curriculum for K-12 Teachers and Hands-on Training Sessions for High School Students

As a way to encourage more (and more diverse students) to consider careers in robotics, faculty members from Bay College and Michigan Tech promotes Robotics Automation to K-12 teachers and high school students. One-day seminar for K-12 teachers is conducted at both Michigan Tech and Bay College in Years 2 and 3 of the project. During the seminar, participants (1) learn concepts of industrial robotics; (2) learn the basics of programming FANUC industrial robots; (3) try the robotic software "RobotRun"; and (4) work with faculty to consider ways the software can be integrated into the K-12 curriculum. Participating teachers are provided with 4 h of theory and 4 h of hands-on operating and programming with FANUC robots and the "RobotRun" simulation software. To promote the field of robotics directly among the high school students, one "day camp" is conducted yearly, at both institutions. Students (1) learn basic principles of industrial robots; (2) operate and program FANUC industrial robots; (3) utilize the gaming environment of the "RobotRun" simulation software to play embedded games and conduct basic programming tasks (in Year 2 and 3). Due to the remote location of Upper Peninsula schools, very few programs targeting STEM fields are available. The proposed camps will provide high school students with the extraordinary opportunity to learn and be engaged in STEM-related activities using the appealing nature of robotics. This early-age engagement in STEM activities will help to create a clear path for the students to continue education through postsecondary institutions.

7.9   Task 4: Robotic Automation Laboratories at Michigan Tech and Bay College

The School of Technology at Michigan Tech offers high-quality, up-to-date academic programs aimed at meeting the immediate and future needs of industry. As a technology program, we offer significant hands-on laboratory experiences and applied research opportunities to undergraduates. These experiences complement the classroom experience and prepare our students for careers in a wide range of industries.

The EET program at Michigan Tech has a state-of-the-art Robotic Automation laboratory, which allows faculty to provide students with training that meets industrial standards. Currently, the laboratory is equipped with four FANUC LR Mate Education Training Carts and incorporates FANUC Robotics' latest-generation electric, servo-driven mini robot, housed in a self-contained, portable enclosure. The LR Mate Education Training Cart MH1 can be used to teach students how to program a real robot, in real time, in a safe, controlled environment. The FANUC robot is a highly upgradable system and allows demonstrating basic functions (e.g., vision, collision guard, path tracing, insert, straight-line accuracy), as well as creating more advanced hands-on laboratories. While it is desirable to have more robots, the four units currently installed at Michigan Tech are sufficient to teach a class of 16 students with two laboratory sections.

Current robot capacity is adequate to train and certify at least eight industry representatives, students from other institutions, or displaced workers. This capability allows for program sustainability and expansion. Training revenue will be used to acquire new robots to further support program growth. Bay College robotic laboratory is equipped with three FANUC LR Mate carts which allows Bay for: (1) introducing two Robotics Automation courses in the curriculum; (2) providing hands-on training to Bay students; (3) training and certifying students from other institutions, industry representatives, and displaced workers; (4) conducting seminars for K-12 teachers; and (5) conducting outreach activities for the high school students.

7.10   Sustainability of the Project and Its Broader Impact

The four certification programs in Industrial Automation: Handling Tool Operation and Programming, Roboguide, Robot Operations, and iR-Vision 2D, implemented at Michigan Tech and Bay College through this project, will attract prospective students, industry representatives, and unemployed workers who want to retool their skills, and students from other universities and colleges without a certification program. The revenue received from these certification programs will serve as a main source of funds to sustain the project and to continually enhance and update the Robotics Automation programs at the partnering institutions. Due to the rapidly evolving technological world, Robotics Automation is currently developing at a fast pace. This pace will only increase in the near future. As a result, the demand for technologists in the field of robotics is also increasing. Educational units must support this growing demand for highly knowledgeable technologists from the industrial sector and, in particular, by technology programs which place an emphasis on hands-on training. To build highly effective and self-sustaining programs with broad impacts in Robotics Automation is not a simple task. Bay College and Michigan Tech have joined efforts to build this program and to make it highly adaptable by various institutions and with different budgets. The curriculum developed in this project and open-source training software "RobotRun" will enable three modes of adaptation, which are shown in Table 7.1.

Table 7.1 Modes of adaptation by other institutions

Modes	Institution budget	Adapted project materials, hardware, and software	Benefits
Mode 1	High	1. All course materials 2. 3–4 Fanuc Industrial Rabun 3. FANVC Roboguide Software 4. Project developed RobotRun Software	1. Teach robotics courses and certify students 2. Provide bands-art training using industrial robots 3. Train students on industrial and educational software packages 4. Train and certify significant number of students from other institutions, industry representatives, and displaced workers. High profit and possibility for fast expansion 5. K-12 outreach activities
Mode 2	Medium	1. All course materials 2. 1–2 Fanuc Industrial Robots 3. FANUC Roboguide Software 4. Project developed Robot Run Software	1. Tench Robotics courses and certify students 2. Provide hands-on training using industrial robots 3. Train students on industrial and educational software packages 4. Train and certify average number students from other institutions, industry representatives, and displaced workers. Medium profit and possibility for slow expansion 5. K-12 outreach activities
Mode 3	Low	1. All course materials 2. Project developed RobotRun Software	1. Teach Robotics courses 2. Train students on educational software package 3. K-12 outreach activities

All three modes will allow any institution to teach robotics skills; modes one and two will also allow for industrial training and certification, which will enable the other new programs to grow and expand.

FANUC is represented in five continents and >22 countries with more than 100,000 robots installed in the USA and 250,000 robots worldwide. The extensive presence of FANUC robots in industry requires well-trained and certified specialists to design, operate, and maintain the robots. College graduates equipped with these skills will be highly marketable, which in turns makes adaptation and implementation of the proposed curriculum in Robotic Automation very attractive for others. These factors suggest that institutions that make the initial investment to implement our curriculum will likely generate a significant revenue stream to not only support but also further expand their programs in Robotics Automation.

7.11   Conclusion

The overall goal of the collaborative project between Michigan Tech and Bay de Noc Community College is to help meet the nation's forthcoming need for highly trained Industrial Robotics workers. Strategies include developing, testing, and disseminating an updated, model curriculum, laboratory resources, and simulation software package suitable for use in both 2- and 4-year EET programs. To complement this effort, outreach to K-12 students and teachers will work to enlarge the pipeline and diversity of students interested in careers in robotics. Programs will also be offered to students at other institutions and to workers in industry to broaden impact.

Described curriculum development not only is geared toward students enrolled in the university program but also provides opportunity for industry representatives to retool their skills in robotics and automation. Developed course and its derivatives are designed to provide significant hands-on training in robotic vision systems and teach skills that are very relevant to current industry needs.

Acknowledgements This project is supported by National Science Foundation grant DUE-1501335.

References

Alimisis, D. (2005). Technical school students design and develop robotic gear-based constructions for the transmission of motion. Warsaw: Eurologo.

Apple investing record $10.5 billion in supply chain robots & machinery. (2016).

Barker, B., & Ansorge, J. (2007). Robotics as means to increase achievement scores in an informal learning environment. *Journal of Research on Technology in Education, 39*(3), 229–243. doi:10.1080/15391523.2007.10782481.

Barnes, D. (2002). Teaching introductory Java through LEGO MINDSTORMS models. *SIGCSE Bulletin, 34*(1), 147. doi:10.1145/563517.563397.

Beer, R., Chiel, H., & Drushel, R. (1999). Using autonomous robotics to teach science and engineering. *Communications of the ACM, 42*(6), 85–92. doi:10.1145/303849.303866.

Chang, D. (2009). Educating generation Y in robotics. Proceedings of ASEE AC 2009-750.

Ciaraldi, M. (2009a). Designing an undergraduate robotics engineering curriculum: Unified robotics I and II. ASEE.

Ciaraldi, M. (2009b). Robotics engineering: A new discipline for a new century. ASEE.

Devine, K. (2009a). Agile robotic work cells for teaching manufacturing engineering. ASEE.

Devine, K. (2009b). Integrating robot simulation and off-line programming into an industrial robotics course.

Downloads—World Robotics 2014. (2016). Worldrobotics.org. Retrieved 30 July 2016, from http://www.worldrobotics.org/downloads

Eslami, A. (2009). A remote-access robotics and PLC laboratory for distance learning program. ASEE.

FANUC Robotics system R-30iA controller iR-Vision with error proofing student (Manual) (1st ed.).

F.A.S.T. (2014). Robot.fanucamerica.com. Retrieved 30 July 2016, from http://robot.fanucamerica.com/support-services/robotics-training/schools.aspx

Fernandez, K. (2009). NASA summer robotics interns perform simulation of robotics technology. ASEE.

Guiding the Independent Learner in Web-Based Training. (1999). *Educational Technology, 39*(3).

International Federation of Robotics: Metra Martech Study on Robotics. (2016). Retrieved 29 July 2016, from http://www.ifr.org/uploads/media/Metra_Martech_Study_on_robots_02.pdf

Johnson, J. (2012). Children, robotics, and education. *IEEE Artificial Life and Robotics, 7,* 16–21.

Karatrantou, A. & Tzimogiannis, A. (2005). Introduction in basic principles and programming structures using the robotic constructions LEGO Mindstorms. In 3rd National Conference, Teaching Informatics, University of Peloponnese.

Liu, Y. (2009). From handy board to VEX: The evolution of a junior-level robotics laboratory course. ASEE.

Mauch, E. (2001). Using technological innovation to improve the problem-solving skills of middle school students: Educators' experiences with the lego mindstorms robotic invention system. *The Clearing House: A Journal of Educational Strategies, Issues and Ideas, 74*(4), 211–213. doi:10.1080/00098650109599193.

Michalson, W. (2009). Balancing breadth and depth in engineering education: Unified robotics III and IV. ASEE.

Morey, B. (2007). Robotics seeks its role in aerospace. *Manufacturing Engineering, 139*(4).

Nourbakhsh, I., Crowley, K., Bhave, A., Hamner, E., Hsiu, T., Perez-Bergquist, A., et al. (2005). The robotic autonomy mobile robotics course: Robot design, curriculum design and educational assessment. *Autonomous Robots, 18*(1), 103–127. doi:10.1023/b:auro.0000047303.20624.02.

Office of Science and Technology Policy, Domestic Policy Council. (2006). American competitiveness initiative—Leading the world in innovation.

Papert, S. (1980). *Mindstorms*. New York: Basic Books.

Piaget, J., & Piaget, J. (1973). *To understand is to invent*. New York: Grossman Publishers.

Ren, P. (2009). Bridjing theory and practice in a senior-level robotics course for mechanical and electrical engineers. ASEE.

Robinson, M. (2005). Robotics-driven activities: Can they improve middle school science learning? *Bulletin of Science, Technology & Society, 25*(1), 73–84. doi:10.1177/0270467604271244.

Robots: More capable, Still Flexible. (2005). *Manufacturing Engineering, 134*(5).

Rogers, C., & Portsmore, M. (2004). Bringing engineering to elementary school. *Journal of STEM Education, 5,* 17–28.

Schneider, R. (2005). Robotic automation can cut costs. *Manufacturing Engineering, 135*(6).

Stienecker, A. (2008). Applied industrial robotics: A paradigm shift. ASEE.

Tolinski, R. (2006). Robots step up to machining. *Manufacturing Engineering, 137*(3).

Witeman, E. & Witherspoon, L. (2003). Using Legos to interest high school students and improve K-12 STEM education. *Frontiers in Education, 2*, 5–8.

You, Y. (2009). A project-oriented approach in teaching robotics application engineering. ASEE.

Chapter 8
STEM Education by Exploring Robotics

Francis Tuluri

Abstract The chapter on 'STEM Education by Exploring Robotics (SEEbots)' describes a wide range of educational robotics modules for increasing STEM learning at pre-college level and at the college level. The recent advances in electronics technology and computer technology are making a variety of novel and versatile robotics modules within the budget of educational institutions for creating interest among the students toward STEM learning, and to broaden participation of students seeking STEM careers. The suggested robotics modules are based on the experiences gained in teaching robotics for STEM education to middle school level to college level, through projects sponsored by funding agencies. The information described is intended for educators as a reference guide to designing their robotics courses for STEM learning in secondary or post-secondary introductory robotics educational program.

Keywords STEM · Robotics · Lego · Technology · Learning · Education

8.1 Introduction

In particular, the chapter discusses some of the topics on introductory educational robotics course entitled 'Introduction to Robotics' taught to the full-time undergraduate students and to K12 students in outreach pre-engineering summer programs. The course is designed as providing transition in robotics education to K12 students entering into higher education with advanced topics on hardware and software of educational and industrial robotics. Similar courses will sustain the skills that the K12 students had in robotics while they begin their journey through higher learning as a passage to seeking STEM career in industry. The goal of the project is to use robotics to increase STEM learning among students through hands-on activities by integrating science, mathematics, computer programming,

F. Tuluri (✉)
Jackson State University, Jackson, MS, USA
e-mail: francis.tuluri@jsums.edu

© Springer International Publishing AG 2017   195
M.S. Khine (ed.), *Robotics in STEM Education*,
DOI 10.1007/978-3-319-57786-9_8

engineering, and technology. The instructional methodologies explored a variety of robotic platforms ranging from educational robot modules (EDR) to industrial robot trainers. To introduce robotics to the entry-level students, three educational platforms are used—Lego NXT/EV3 robot, VEX cortex robot, and Parallax Beebot. A FANUC robot trainer is also used for teaching skills on programming industrial robot. For programming practice on EDRs, graphical language and higher level language (C language) are used. For programming robo arm trainer, propriety software loaded on its teach pendant is used. Additionally, a simulation software package (Handling-Pro) is also used to strengthen practicing programming industrial robots. See additional resources given at the end of the chapter for information on Lego, VEX, Parallax, and FANUC brands. The information provided in this chapter is intended for use by educators to introducing robotics as a learning platform in STEM education in schools, community colleges, and universities. The course content is developed from the projects sponsored by funding agencies—Title III DoE grant, Army Educational Outreach Program (UNITE and REAP/USAEOP), and Lockheed Martin Foundation.

8.2 Course Overview

The 'Introduction to Robotics' course is targeted to regular freshman undergraduates or K12 high school students interested in taking summer pre-engineering program with a view to broadening STEM participation. The course provides 3 credits to the freshman undergraduates and is given in every spring semester. The pre-engineering summer program is offered for four weeks during the month of June. In addition to hands-on activities on programming educational robotics modules, the course also covers basics of topics such as sensors, digital automation systems, fluid automation systems, pneumatic automation systems, and industrial robotics. Upon competition of the course, the student will have the ability to:

- describe the distinction between educational robotics and industrial robotics
- build educational robotics modules
- program educational robotics modules
- review electronics (analog and digital)
- understand the function of sensors and actuators
- navigate robotics modules using sensors
- explain automation systems—digital, electromechanical, fluid, and pneumatic systems
- state and explain the functions of industrial robot
- select and apply a knowledge of mathematics, science, engineering, and technology to engineering technology problems that require the applications of principles and applied procedures or methodologies
- identify, analyze, and solve broadly defined engineering technology problems;

- apply written, oral, and graphical communication in both technical and non-technical environments, and an ability to identify and use appropriate technical literature.

8.3 Motivation

In traditional teaching methods of science and engineering, students lack experience of applying physical principles to real-time physical situation required in these disciplines of study (Barreto and Benitti 2012). Students are not properly engaged and inspired in learning through interactive activities. Lack of interest combined with shortage of practical experience in learning eventually leads them to failure to keep up with the grades expected of them or even can lead to increased rates of attrition in science and engineering disciplines (Graham et al. 2013). Robotics is an interactive discipline through which the learner can get the feeling of learning by achievement through designing, building, programming, and executing (Rockland et al. 2010; Mataric et al. 2007; Nugent et al. 2010; Wang 2016). Additionally, robotics integrates the application of other disciplines such as science, mathematics, electronics, and technology. Robotics educational modules enable the students learn principles of science and mathematics by working through real-world applications by using system integrated approach (Barker 2012; Sullivana and Heffernana 2016; Eguchi 2016).

Further, recent advances in technology are providing EDRs to classroom instruction at a very modest and affordable price with excellent computing features. EDRs come up with immense power for use in education as an interactive and intelligent tool to study variety of physical systems through sensors. The electronics system facilitates collecting the physical data which can be used to study the behavior of the physical system of interest (Teslya and Savosin 2014).

Robotics education creates interest to the students to get started with learning big ideas in STEM education through experiencing practical applications that go beyond classroom learning. Further, the students are engaged and inspired to be prepared for a fruitful career in STEM-related workforce. With recent advances in technology, robotics industry is on the rise leading to robots being used in every walk of life—from domestic appliances, to manufacturing, to automation, and to military. As the competition in industry is getting critical, the students have to face challenges to procure multiple skills. Robotics education is the means to make the students of K12 level and beginner undergraduate level excel in STEM disciplines to aspire related careers.

8.4   Lego NXT/EV3 Platform

Lego Mindstorms educational kit (Wang 2016) contains a microcomputer (NXT or EV3), lego elements (axles, connectors, wheels, gears, etc.), sensors, and servo-motors required for students to assemble and program a robot. The kit comes with basic sensors such as touch, sound, ultrasonic, light, and gyro; however, a variety of sensors are also available from a third party. See Fig. 8.1 for a robot built with Lego NXT kit and Tetrix EV3 kit.

The microcomputer can be programmed using a software package called NXT G (a graphical programming) or RobotC (a high-level C-programming) (Cruz-Martin et al. 2012). Graphical programming is more appropriate to introduce programming concepts for a beginner-level student in robotics education (Grandi et al. 2014). While the Lego kit is popularly used in robotic completions at K12 level (Grandi et al. 2014), it can also be extended to design and study a programmable physical system (Cruz-Martin et al. 2012). The configuration of the programmable physical system consists of five stages, namely (1) building the robot, (2) connecting the robot sensors, (3) programming the robot to collect data, (4) analyzing, and (5) interpreting the results to study the behavior of the system.

'Introduction to Programming EV3' curriculum of Lego provides a series of topics for getting acquainted with basic robotics and programming for all levels of students. Here, we describe an approach with a few examples of using educational robotics to further STEM learning by designing novel programmable physical systems.

(1) 1 D motion under uniform acceleration

To study one-dimensional motion of an object under constant acceleration, a physical system can be designed with a NXT/EV3 mounted over a trolley (a cart, for example, with freely rotating wheels) rolling down an inclined flat surface (for example, a wooden plank) (Tuluri et al. 2014). A light sensor is attached to the robot module. The trolley is made to move on a wooden plank with ten strips of

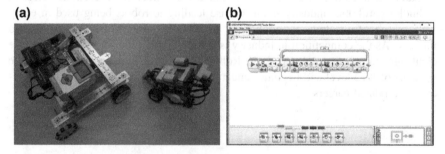

(a)   **(b)**

Fig. 8.1 An educational robot module of Lego NXT, and EV3 on Tetrix platform (**a**) An example of EV3 graphical programming work space (**b**)

paper equally spaced 5 cm apart, along its length (see Fig. 8.2). The robot module is mounted on a trolley with its light sensor moving over the strips of paper.

The robot is released from rest on the inclined plane. Then, the robot module is programmed in remote data logging mode to record the changes in the reflectivity of the light sensor with time as it moves past the stripes of paper. The reflectivity markers enable to locate the position of the robot as it rolling down the inclined plane. Let v is the initial velocity of the cart, θ is the angle of the incline, and g is the acceleration due to gravity. By simple physics principles (Halliday et al. 2014a), the acceleration of the moving robo-trolley on the inclined surface can be obtained from the study of distance versus time^2 plot relationship given by,

$$ s = \left(\frac{1}{2}\right)(g \sin \theta)t^2 \tag{8.1} $$

where s is the distance, t is time, and g is acceleration due to gravity.

Since the physical system is an intelligent and programmable, the students can extend the concept to predict and test hypothesis. For example, the experiment can be repeated for measuring the acceleration due to gravity for different slopes of the inclined surface and extrapolate to find the acceleration due to gravity in free fall, or the effect of friction on the motion of the trolley can also be estimated and compared with theoretical expectations.

(2) RC constant

In a series resistor (R) and capacitor (C) electric circuit, measurement of RC time constant is a simple activity for demonstrating the working principle of a capacitor (Halliday et al. 2014b; Floyd and Buchla 2010a). The experiment relates principles of electronics coupled with mathematical concept of exponential growth or decay which is common in many real-time situations of science and engineering. The time constant of a series RC circuit describes the rate of charging/discharging of the capacitor and is mathematically given by,

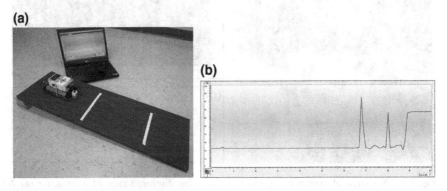

Fig. 8.2 A physical system of robo-trolley on an inclined plane for measuring acceleration of the moving cart (**a**) A sample of photo sensor peaks as the cart is moving past the paper strips (**b**)

$$Timeconstant, \tau = R * C \qquad (8.2)$$

where R is the resistance and C is the capacitance of the series RC circuit.

Using educational robotics for measuring RC time constant enables students to learn through designing a programmable physical system. In this experiment, a simple RC circuit is built on a breadboard with a two-way switch for charging and discharging the capacitor. The capacitor is charged from a 9 V battery and is disconnected from the source to discharge through a LED. The light sensor of the robo module is programmed in data logging mode (remote or wired) to record the decreasing intensity of the LED (Tuluri et al. 2014; Tuluri 2015) (See Fig. 8.3). The experiment can be explored for different resistor and capacitor combination to examine the rate of decay plots and compared with theoretical predictions.

(3) ADC

Analog-to-digital conversion is a commonly used principle in all digital processing systems. The touch sensor of a Lego NXT or EV3 microcomputer can be tweaked and can be used to teach and study analog-to-digital conversion (Tuluri 2015). Inside the microcomputer circuitry, the touch sensor port is basically a part of a voltage divider (Floyd and Buchla 2010b), and analog voltage signal across the sensor is digitized through an analog-to-digital converter (ADC) (Floyd 2006). The voltage divider is powered by 5 V supply line of the microcomputer and consists of a fixed 10 K Ω resistor and the sensor resistance. The sensor resistance is measured between the pin#1 of the sensor cable (white color) and pin#2 (black color) which is connected to the ground (zero volts). Thus, a resistance sensitive probe is made by cutting the sensor end of the cable and bringing out the terminals of pin 1 and pin 2 (see Fig. 8.4a). Corresponding to the change in the sensor resistance between zero (short) and infinite (open), the ADC measures the voltage between 5 and 0 V in intervals of 1023 (2^{10} bits) with a resolution of 4.88 mV (Floyd 2006). Using the

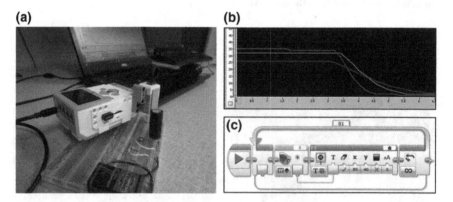

Fig. 8.3 EV3 physical system arrangement for studying discharge rates of a capacitor (a) A sample of data logging plot using NXT module (b) A EV3 program to display sensor value on the module LCD (c)

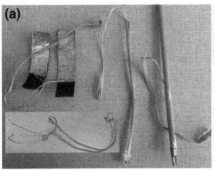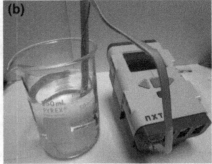

Fig. 8.4 A sample of resistance sensitive probes for studying physical systems (**a**) NXT physical system set up for demonstrating ADC principle (**b**)

principles of voltage divider and ADC, the resistance of the resistive sensor probe (R) can be expressed in terms of the raw value of the ADC (Raw) as,

$$R = \frac{10,000 * \text{Raw}}{1023 + \text{Raw}} \, \Omega \qquad (8.3)$$

Several types of resistance sensitive probes can be designed based on the study of interest (See Fig. 8.4a). A programmable physical system can be designed to study physical quantities of interest. For example, the open ends of the resistance sensitive probe can be immersed in fluids to study humidity (See Fig. 8.4b), or can be connected to a thermistor to study temperature variation, or as a skin resistance sensor value. A simple program will enable to display the resistance of the resistance sensitive probe (See Fig. 8.5).

8.5 VEX Cortex Platform

Robotic competitions are becoming very popular method of inspiring the students toward learning by experiences (Caro 2011; Nugenta et al. 2016). In view of the availability of educational resources, many institutions are adopting VEX robotics into their curriculum to increase STEM learning (Sullivan et al. 2016). Further VEX robotics modules have rigid framework design and powerful processor and are also gaining popularity in robotic competitions at K12 and undergraduate level to broaden STEM participation. We have used VEX cortex robotic educational module to develop working knowledge of computer hardware and software and teach problem solving skills in STEM education. We have used RobotC to teach advanced skills in programming the robot.

VEX cortex enables to set the speed of the wheels independent of the other—the magnitude and direction. Here, we show one example of combining kinematics

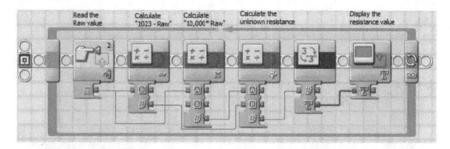

Fig. 8.5 NXT program to display the resistor sensitive probe value on the module display panel

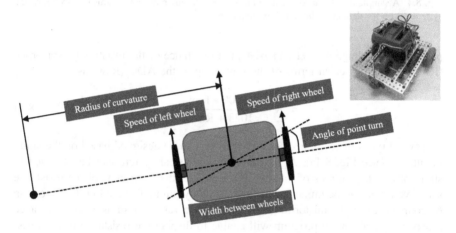

Fig. 8.6 Geometrical formulation for angular rotation or spin rotation of a robot. The inset (on the *top-right corner*) shows an assembled VEX cortex educational robot

principles with mathematics in the context of navigating the motion of a robot along arc of a circle of certain radius of curvature (see Fig. 8.6).

Consider that 'd' is the diameter of the wheels, b is the center-to-center distance between the drive wheels, and R is the radius of curvature of the robot's turn. Let V_R and V_L are the speeds of the right and left wheels, respectively. Using the basic principles of angular velocity, one can calculate the radius of curvature of robot's turn in terms of V_R and V_L (Halliday et al. 2014c) given by,

$$R = \left(\frac{b}{2}\right) \frac{(V_R + V_L)}{(V_R - V_L)} \tag{8.4}$$

The sign of R determines the direction of rotation. As a special case, if the speed of the wheels is equal then the robot moves linearly corresponding to infinite radius of curvature. The other cases are pivot rotation (keeping the speed of one of the wheels zero) and point rotation (keeping the speed of the wheels equal and

```
#pragma config(Motor,  port2,        leftMotor,   tmotorNormal, openLoop,
driveLeft)
#pragma config(Motor,  port3,        rightMotor,  tmotorNormal, openLoop,
reversed, driveRight)
//*!!Code automatically generated by 'ROBOTC' configuration wizard        !!*//

void forward()
{
        wait1Msec(2000); // Robot waits for 2000 milliseconds before executing
program
        // Move forward at full power for 3 seconds
        motor[rightMotor] = 63;            // Motor on port2 is run at 63 power
forward
        motor[leftMotor]  = 63;            // Motor on port3 is run at 63 power
forward
        wait1Msec(3000);
        }
void turn()
{

        // turn for 1.5 for 3 seconds
        motor[rightMotor] = 20;            // Motor on port2 is run at 20 power
forward
        motor[leftMotor]  = 50;            // Motor on port3 is run at 50 power
forward
        wait1Msec(1500);
        }
task main()
{

        forward();
        turn();
        stopAllMotors();
}
```

Fig. 8.7 An example of RobotC programming for a robot moving in a curve linear motion

opposite) which are useful for programming the robot for making turns in a labyrinth. An example of RobotC programming for a robot moving in a orbital turn is given in Fig. 8.7.

Using the mathematical solution, one can estimate to program a robot move in a complex path (see Fig. 8.8). Similar examples facilitate the students toward problem-based learning investigation. Working on such hands-on activities increases student's skills in problem solving, and critical thinking and engaging them through predicting, and validating the principles with measuring.

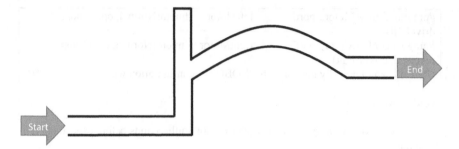

Fig. 8.8 An example of navigating a robot in straight segments and turns in a complex track of motion

Fig. 8.9 An assembled circuitry of sensors on Boe-Bot for programming

8.6 Parallax Stamp 2 Platform

We have used Parallax Boe-Bot robot to teach working principles of sensors and controlling them through a microcontroller. Boe-Bot kit is equipped with Basic Stamp II microcontroller and comes with servomotors for locomotion and a variety of sensors. Students are also introduced to build circuits on the onboard breadboard and program and control the sensors. Further, students with little or no programming experience can easily program the microcontroller using PBASIC—a BASIC programming version for the Stamp II controller. A set of simple hands-on activities is designed for the students to learn the working principles of basic sensors such as photoresistor, ultrasonics sensor, IR sensor (transmitter and receiver), piezoelectric force sensor. A sample of the assembled circuit of the robot with IR and ultrasonic sensors to detect an obstacle and avoid collision in an autonomous mode is shown in Fig. 8.9.

8.7 FANUC Robot Trainer

Because of the growing demand of robotics automation in manufacturing and the importance of interdisciplinary STEM skills in the workforce, we have used a few topics in our robotics course on FANUC Robotics LRMate Education Training Cart MH1 (see Fig. 8.10).

Simple hands-on activities on using industrial robot trainer help the students to prepare them to moving from educational robotics to industrial robotics. These activities will also motivate the K12 and undergraduate students to aim for successful career in engineering in particular and robotics automation in general. Motivating the students on bridging the skills gap between industry and academia is essential to prepare the students in meeting challenges in the high-tech-oriented workplace of the global market. Particularly, when students are exposed in the classroom to real-world industrial equipment and environment that they are going to be working on, they will be engaged and focused on learning STEM disciplines.

In our course on 'Introduction to Robotics,' we have used Robo guide and used Handling-Pro to teach topics such as frames of reference. The students are given hands-on activities to use Teach Pendant to make the arm of the robot move in joint frame reference, user frame reference, and world frame reference. An example of the robo arm moving along the sides of a rectangle in user-defined coordinate frame of reference is shown in Fig. 8.11. Similar activities were also done in robotic simulations using the Handling-Pro. See Fig. 8.12 for a simple pick and drop example in robotic simulations, and Fig. 8.13 for instructions available on the help menu of the software.

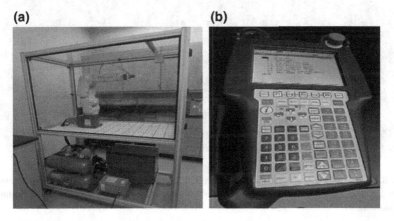

(a)   **(b)**

Fig. 8.10 FANUC Robotics LRMate Education Training Cart MH1 (**a**) and the teach pendant (**b**) for programming and teaching the robo arm

(a)   (b)

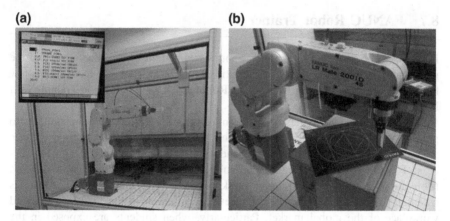

Fig. 8.11 Programming Robo Arm in a user-defined frame of reference along the sides of a square (**a**) is the home position of the robo arm (the inset on the *top-left corner* is the program code on the Teach Pendant) and (**b**) is a snapshot of the robo arm during the execution of the program in auto mode

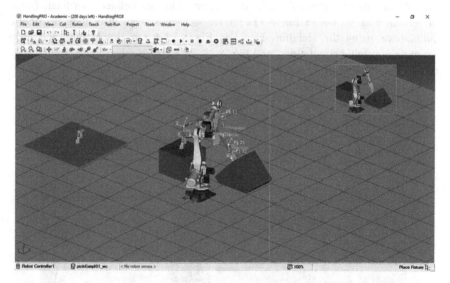

Fig. 8.12 A snapshot of Pick and place robotic Simulations using Handlin-Pro simulation software. The inset (on the *top-right corner*) shows the instance of picking the object during the simulation

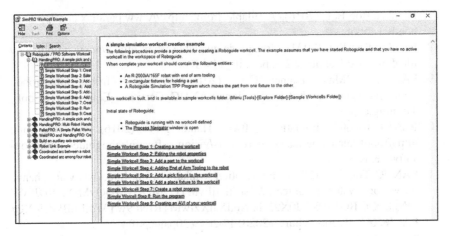

Fig. 8.13 Handling-Pro easy to do instruction for writing programming for simple examples

8.8   Conclusion

An approach to broadening STEM participation using robotics education is described in the chapter. The methodology is based on a course on an introductory robotics course to engage and inspire entry-level undergraduate students (of any discipline) and K12 students for pursuing STEM education. The examples covered are to showcase how the students can be challenged to develop interdisciplinary skills, problem solving skills, reasoning skills toward attaining critical thinking in their studies of interest through tactile experience of real-world environment. The information can be extended by users of their interest and is also intended for educators as a reference guide for designing their course in secondary or post-secondary robotics program.

Additional Resources

1. Lego Mindstorms. (2016). Available from: http://www.lego.com/en-us/mindstorms/ ?domainredir=mindstorms.lego.com
2. Lego EV3. (2016). Downloads available from: http://www.lego.com/en-us/ mindstorms/downloads
3. Lego EV3, (2013). User Guide available from: https://le-www-live-s.legocdn. com/sc/media/files/user-guides/ev3/ev3_user_guide_en-20dff91180424fe8a914 8052f5288c82.pdf
4. Lego NXT, (2009). User Guide available from: http://lego.brandls.info/ebooks/ 8547_ms_user_guide.pdf
5. VEX Cortex EDR. (2015). Curriculum available from: http://curriculum. vexrobotics.com/curriculum, http://curriculum.vexrobotics.com/home
6. VEX Cortex EDR. (2015). Teacher resources available from: http://www. education.rec.ri.cmu.edu/robots/vexteacher/cortex/D

7. Parallax Boe-Bot. (2016). Available from: https://www.parallax.com/product/boe-bot-robot
8. Parallax Boe-Bot. (2004). User guide available from: https://www.pololu.com/file/download/Roboticsv2_2.pdf?file_id=0J208
9. FANUC LRMate Trainer Cart. (2016). Available from: http://robot.fanucamerica.com/robot-applications/FANUC_Certified_Education_Robot_Training.aspx
10. FANUC robotics simulations software Handling-Pro. (2016). Available from: http://robot.fanucamerica.com/products/vision-software/roboguide-simulation-software.aspx
11. FANUC Training. (2016). Information available from: http://www.alleghenyedusys.com/system/resources/W1siZiIsIjIwMTQvMDQvMDQvMDhfNDJfMTFfNzIxX0ZBQ19DZXJ0X2Jyb2NodXJlXzIwMTRfNF9wYWdlLnBkZiJdXQ/FAC%20Cert%20brochure%202014%204%20page.pdf

References

Barker, B (Ed). (2012). *Robots in K-12 education: A new technology for learning: A new technology for learning.* ISBN 978-1-4666-0182-6, IGI-global.

Barreto, F., & Benitti, V. (2012). 2012. *Exploring the educational potential of robotics in schools: A systematic review, Computers & Education, 58,* 978–988.

Caro, A.I. (2011). VEX robotics: STEM program and robotics competition expansion into Europe. In D. Obdrzalek & A. Gottscheber (Eds.), *Research and education in robotics—EUROBOT* (pp. 10–16). Available from: http://ieeexplore.ieee.org/stamp/stamp.jsp?tp=&arnumber=7000936

Cruz-Martín, A., Fernández-Madrigal, J. A., Galindo, C., González-Jiménez, J., Stockmans-Daou, C., & Blanco-Claraco, J. L. (2012). A lego mindstorms nxt approach for teaching at data acquisition, control systems engineering and real-time systems undergraduate courses. *Computers & education, 59,* 974–988.

Eguchi, A. (2016). Computational thinking with educational robotics. In *Proceedings of Society for Information Technology & Teacher Education International Conference 2016* (pp. 79–84). Association for the Advancement of Computing in Education (AACE).

Floyd, T. L. (2006). *Digital fundamentals* (9 ed., pp. 744–754). Pearson—Prentice Hall, ISBN-0-13-194609-9.

Floyd, T.L., & Buchla, D.M. (2010a). *Electronics fundamentals—circuits, devices, and applications* (pp. 409–414). Prentice Hall, ISBN-13: 978-0-13-507295-0.

Floyd, T. L., & Buchla, D. M. (2010b) *Electronics fundamentals—circuits, devices, and applications* (pp. 135–141). Pearson—Prentice Hall, ISBN-13: 978-0-13-507295-0.

Graham, M. J., Frederick, J., Byars-Winston, A., Hunter, A., & Handelsman, J. (2013) Increasing persistence of college students in STEM, *Science 27, 341*(6153), 1455–1456 (2013) doi:10.1126/science.1240487

Grandi, R., Falconi, R., & Melchiorri, C. (2014). Robotic competitions: Teaching robotics and real-time programming with LEGO. In *Proceedings of the 19th World Congress: The International Federation of Automatic Control.* South Africa.

Halliday, D., Resnick, R., & Walker, J. (2014a). *Fundamentals of physics* (10th ed., pp. 23–25, and 94–129). Wiley Publications, ISBN-10: 1118230647; ISBN-13: 978-1118230640.

Halliday, D., Resnick, R., & Walker, J. (2014b). *Fundamentals of physics* (10th ed, pp. 788–792). Wiley Publications, ISBN-10: 1118230647; ISBN-13; 978-1118230640.

Halliday, D., Resnick, R., & Walker, J. (2014c) *Fundamentals of physics* (10 ed, pp. 257–270). Wiley Publications, ISBN-10: 1118230647; ISBN-13: 978-1118230640.

Mataric, M.J., Koenig, N., & Feil-Seifer, D. (2007). *Materials for enabling hands-on robotics and STEM education*. American Association for Artificial Intelligence (www.aaai.org). Available from: http://www.aaai.org/Papers/Symposia/Spring/2007/SS-07-09/SS07-09-022.pdf

Nugent, G., Barker, B., Grandgenett, N., & Adarnchuk, V. I. (2010). Impact of robotics and geospatial technology interventions on youth STEM learning and attitudes. *Journal of Research on Technology in Education, 42*(4), 391–408.

Nugenta, G., Barkera, B., Grandgenett, B., & Welcha, G. (2016), Robotics camps, clubs, and competitions: Results from a US robotics project. *Robotics and Autonomous Systems, 75*(B), 686–691.

Rockland, R., Bloom, D. S., Carpinelli, J., Burr-Alexander, L., Hirsch, L. S., & Kimmel, H. (2010). Advancing the "E" in K-12 STEM education. *The Journal of Technology Studies, 36* (1), 53–64.

Sullivana, F., & Heffernana, J. (2016). Robotic construction kits as computational manipulatives for learning in the STEM disciplines. *Journal of Research on Technology in Education, 48*(2), 105–128. doi:10.1080/15391523.2016.1146563

Teslya, N., & Savosin, S. (2014). Smart-M3-based robot interaction in cyber-physical systems. In *Proceeding of the 16th Conference of FRUCT Association, 2014*, pp. 108–114.

Tuluri, F. (2015). Using robotics educational module as an interactive STEM learning platform. In *ISEC (IEEE), 5th IEEE Integrated STEM Conference, Proceedings, 2015*, pp. 16–20.

Tuluri, F., Colonias, J., Vance, D., Dixon, D., White, M., & Edwards, A. (2014). Robotics-based educational tool—An interactive learning platform to enhance understanding behavior of physical systems. *The Researcher, 27*(1), 89–104.

Wang, W. (2016). A mini experiment of offering STEM education to several age groups through the use of robots. In *2016 IEEE Integrated STEM Education Conference (ISEC)*, pp. 120–127.

Part III
Robotics, Creativity and STEAM Education

Chapter 9
The Creative Nature of Robotics Activity: Design and Problem Solving

Florence R. Sullivan

Abstract Learning with robotics through open-ended design challenges enables the development of creativity in children. Such robotics learning environments encourage three creative activities that are closely identified with problem solving and design activity, including problem finding, idea generation, and invented strategies. In this chapter, I present the theoretical grounding for the relationship of creativity to robotics, with a special emphasis on the role of play in robotics learning. I provide examples, drawn from my research, to demonstrate how children engage in play and creativity with robotics, and I discuss how these activities relate to learning. In addition, I provide curricular and pedagogical recommendations to teachers interested in supporting student creativity through a robotics learning unit. Curricular recommendations focus on the types of open-ended challenges teachers may enact across the STEM disciplines. Pedagogical recommendations detail a progressive approach to teaching that emphasizes the facilitating role of students' interest, students' experience, and students' collaborative interactions in learning.

Keywords Robotics · Creativity · Play · Design · Problem solving

9.1 Introduction

Robotics learning revolves around two activities that are creative in nature: design and problem solving. When students work on a robotics challenge, they are involved in the design of a system that includes the building and programming of a robotic device. Designing such a system requires creativity. Oftentimes, students run into unanticipated problems that they must solve in order to create a well-functioning robotic device. Solving such problems also requires creativity. The STEM-based creative activity of design and problem solving has certain attributes

F.R. Sullivan (✉)
University of Massachusetts, Amherst, MA, USA
e-mail: florence@umass.edu

© Springer International Publishing AG 2017
M.S. Khine (ed.), *Robotics in STEM Education*,
DOI 10.1007/978-3-319-57786-9_9

that students can manifest in the classroom with the appropriate pedagogical and curricular support. This creative activity of designing and problem solving in a robotics context may lend itself to other, nearby engineering and computing problems, for example, building a stable bridge or similar structure, creating a mobile application, or developing an interactive Web site. In this way, robotics learning is a creative endeavor with wider application across the STEM disciplines.

In this chapter, I introduce a sociocultural definition of creativity as collaborative dialogic inquiry, which is rooted in the work of Bakhtin (1981). I then discuss the role of play in learning with robotics with examples of such, drawn from my research with colleagues. Next, I provide a theoretical view of the creative nature of design and problem-solving activity, I describe empirical findings that support the theory, and finally, I present a discussion of the pedagogical and curricular approach that best supports creative activity.

9.2 Creativity as Collaborative Dialogic Inquiry

Bakhtin (1981) developed a theory of communication and language development called dialogism. Dialogism has three, interrelated components: addressivity and answerability, heteroglossia, and ideological becoming. Addressivity and answerability refer to Bakhtin's (1986) notion that in any communicative act, the speaker's utterance is influenced by whom she is addressing and the anticipated response of the addressee. In other words, every speaker considers who they are speaking to, prior to the act of speaking. Utterances are formulated with a specific audience in mind. The addressee or audience, thereby, helps to shape what the speaker will say. In this way, utterances among individuals are social and multi-voiced in nature, always containing the voice of the speaker and the anticipated answering voice of the addressee.

Heteroglossia refers to the dynamic and constantly evolving nature of language that is a result of the tension created by the many social languages that make up a national language. With any national language, there are always variations in language usage at the level of social groups that exist within the nation. These social groups may develop slightly different definitions of particular words and/or may coin new words that then become part of the national language lexicon. In Bakhtin's (1981) theory, then, the meanings of words are not fixed, but are ever-changing and dependent on the social and/or ideological position of the speaker of the words and the social context in which they are spoken. Bakhtin describes this phenomenon as the centrifugal forces always affecting any language. At the same time, there are constant centripetal forces that seek to fix the meaning of words in order to create a "unitary language." These centripetal forces are, generally speaking, more institutional in nature in that they seek to reify specific meanings in particular, formal ways. The tension created by these opposing social forces also contributes to the dialogic nature of language. Newer meanings and words are created within the context of this social milieu.

Ideological becoming refers to the development of consciousness within any individual. Consciousness in this regard is the development of one's ontology. Bakhtin (1981) argues that it is through participation in and exposure to various social languages within one's lifeworld that one develops notions and ideas about how the world functions. It is the interplay of ideas espoused in various realms (family, friends, school, religious settings, books, T.V., radio, Internet, etc.) that one begins to form one's own ideas about life and the world. In this way, ideological becoming is a creative act that emerges out of the "interanimation" (p. 345) of the utterances/discourses to which one is exposed.

According to Bakhtin (1981), the utterances one is exposed to can be characterized as one of two types of discourses, internally persuasive discourses or authoritative discourses. Internally persuasive discourses are utterances and ideas that are dialogic in nature. In other words, these utterances are always informed by the addressee, and they are open, dynamic, living, responsive utterances. Authoritative discourses, on the other hand, are monologic in nature. These are dogmatic ideas that have been institutionally reified over time, these utterances are not responsive to the audience, but are always the same and purport to represent "the truth,"—these utterances do not admit the voice of the other in formulation, they do not reflect the perspective of the other, only of the hierarchical authority from which they emanate. In this way, authoritative discourses constitute one-way communication, from the speaker to the addressee. Bakhtin considered these monologic, authoritative discourses to be "dead" discourses. Such a discourse may not be entered into or changed in any way, and it must either be accepted or rejected as it is. The clearest example of this type of discourse is religious doctrine and dogma. True believers accept the religious pronouncements as is and attempt to conform themselves to these ideas.

Of these two types of discourses, it is the internally persuasive discourses that account for creativity. As noted above, it is the interplay of various internally persuasive discourses that account for the development of one's consciousness, but, in addition to this, it accounts for the creativity of the individual as manifested in her actions (including her thoughts). Indeed, Bakhtin (1981) argues that "The semantic structure of an internally persuasive discourse is not finite, it is open; in each of the new contexts that dialogize it, this discourse is able to reveal *ever newer ways to mean*" (p. 346) emphasis added. Ever newer ways to mean is reasonably interpreted as the creation of new ideas and the creation of new meanings. In this way, creativity is a fundamental aspect of meaning making and cognitive development.

In Bakhtin's formulation of dialogism, it is clear to see the connection between creativity and learning. Like creativity, learning may also be viewed as a process that unfolds through the understanding of "ever newer ways to mean." Creativity as collaborative dialogic inquiry, then, refers to both the process of meaning making and the making of new meanings through the interanimation of internally persuasive discourses. Again, internally persuasive discourses are the ideas offered from multiple perspectives within a culture or society. An individual takes in these discourses, considers them in light of one another, and in this way, makes sense of the world. And, at times, the individual creates a new meaning that others have not, yet, considered. Hence, creativity and creative ideas emerge from the social, from the juxtaposition of voices in society as represented in internally persuasive discourses.

In a Bakhtinian sense, voice refers not only to the actual voices of actual speakers, but also to voices as they are set on the pages of a book. Bakhtin was a literary theorist who worked to illuminate our understanding of language and society through examination of literary works. According to the theory of dialogism, the author's voice is reified on the page. Importantly, as Kristeva (1986) has pointed out, the author's voice is speaking within the context of other textual voices. In an elaboration of the theory of dialogism, Kristeva argues that every text is created with other texts in mind, and the author is speaking to an audience that is formed, in part, by other authors and other texts. The voice of the author so reified becomes part of a larger, ongoing textual conversation—this phenomenon is known as intertextuality. I argue that this reification of voice takes place not only in a written text, but in the design of tools that people take up and environments they occupy. For example, in the case of robotics construction kits, the voice of the designer is embedded in the design of the tool. Indeed, the designers of the RCX brick, the prototype of all of the LEGO robotics toys, are Mitch Resnick and Fred Martin. In a chapter published in 1991, Resnick and Martin discussed the design of the RCX along the lines of the features of the microcomputer that enable student learning. These features represent the voice, the ideas, of the designers. When children use the robotics kits, they are, in fact, interacting with the voices and ideas of the designers—these reified voices help to shape student activity and learning through the design of the device and what the features of that design makes possible. Norman (2002) calls this the system image. According to Norman, knowledge of how a device works is encoded in the design of the device. A well-designed device suggests how it is to be used—for example, a handle suggests the action of pulling or pushing. The system image is a means of communication between the designer and the user of the device. And, this communication is dialogic in nature, as the designer has well-considered the addressee (the user of the device) when considering how to design it's features.

In the case of robotics, children interact with the voice/ideas of the designers of the device through play. Indeed, the designers explicitly created a device with which children would wish to play. Resnick (2006) has emphasized the role of playful learning in the learning technologies he has been part of creating, which includes the RCX brick (Resnick & Martin 1991), Pico Crickets (Rusk, Resnick, Berg, & Pezalla-Granlund Rusk et al. 2008), and Scratch (Resnick, et al. 2009). This is an important point to understand, by creating a device with which children can play, the designers have both motivated student learning activity and also introduced a child-friendly mode of inquiry that may lead to both creativity and learning. Let us now consider play in greater depth.

9.3   Play

9.3.1   Role of the Object in Play

One very notable aspect of robotics learning activity is the robotic device itself, given its status as both a manipulable and a computational object. The robotic device is similar in size to a smartphone. It is a microcomputer that can understand and execute commands written in various programming languages. One can easily hold the robotic device in one's hand. While it may be slightly unwieldy for young children, children in middle childhood and older youth have no difficulty holding and manipulating the device. Perhaps due to the handiness of the robotic device, children and youth tend to play with it; it has almost a Heideggerian "ready-to-hand" quality in the way that children immediately identify the robotic device as a playful object. Indeed, in my research with colleagues over the last decade, we have repeatedly observed the playful attitude students take toward the device. This playfulness generally manifests in two modes: anthropomorphizing play and analogical play. In the former mode, my research colleagues and I have observed children pick up a robot and make it dance. We have seen children cradle a robotic device like a baby and wonder aloud what to name it. We have heard children scold a robotic device that is not functioning the way the student believes it should (hence the robot is misbehaving) (Sullivan & Wilson 2015). All of this anthropomorphic play is an example of both imagination (what could the robot be?) and investigation (what is the robot?).

In the mode of analogical play, we have observed children pick up the robot and discuss what other objects the robot reminds them of, for example, children have observed that the robot looks like a drag racing car, a robot with cables attached reminded a child of a purse, yet another child imagined the robot as a dog. These interactions with the robot are not only acts of imagination and investigation, but also acts of identification. In one instance, a middle-school-aged female student playing with one of the small LEGO figurines that come with the robotics kits exclaimed "let's make it a girl" (Sullivan & Wilson 2015). In making this suggestion, the female student was identifying her own gender with that of the nondescript LEGO toy; she was creating space for herself in, what many view as, the male-identified activity of robotics (American Association of University Women 2010).

We have also taken note of the role popular culture depictions of robotics play in students' imaginary play in robotics learning settings. Oftentimes, the only prior knowledge children and youth have about robotics is the popular culture depictions they have encountered in television shows or movies. For example, a middle-school-aged female student we worked with at a daylong robotics workshop for girls continually referred to the LEGO robotic device as "Evie," based on the female robot from the Disney/Pixar movie Wall-E, known as Eve. Her reference accomplished two goals, first she was able to playfully identify with the activity through her prior knowledge of the Eve robot character, and second she was able to play with ideas related to the functioning of the robotic device. For example, she discussed tasks she thought her group should program "Evie" to execute.

9.3.2 Play as a Mode of Inquiry in Creative Activity

Both anthropomorphic and analogic plays contain aspects of investigation that lead to more robust, playful inquiry that deals directly with conceptual aspects of robotics learning. For example, I have observed students engage in analogic play with the light sensor in such a way as to improve their understanding of how the light sensor functions. In this study, the students had programmed the light sensor to respond to a light reading that was lower than a certain threshold. In other words, the sensor was triggered to activate a particular programming sequence when it detected a dark colored object. Through play, the students came to realize they could trigger the light sensor with the tip of their black shoes (Sullivan 2011). The students developed the analogy of the light sensor as a magnet, claiming the light sensor was following their shoes (see Fig. 9.1). While this analogy is weak (a light sensor would only act similarly to a magnet if it was programmed to detect a certain reflected reading and to stay on that reading, for example, as in a line following program), it helped the students begin to understand that the light sensor is programmed to respond to certain environmental conditions, which is the conceptual essence of sensors. We have observed that play is an important mode of inquiry for children when they are working with robotics; it stimulates their interest in the topic and allows them to consider robotics in light of prior knowledge; it allows them to identify with the activity (which is particularly important for girls); and it allows them to develop deeper conceptual understanding of the functioning of the robot, for example, the role of sensors in robotics activity.

Having defined creativity from a sociocultural lens and considered play as an important element of student creative learning with robotics, let us now consider the specific creative activities that are most closely related with formal robotics study: design and problem solving.

Fig. 9.1 Students playing with triggering the light sensor with their shoes

9.4   The Creative Nature of Design and Problem-Solving Activity

As noted earlier in the chapter, the activity of robotics is comprised of both engineering design and computer programming elements. Both of these activities are creative by nature. In robotics, an important aspect of designing the robot and writing the computer program is problem solving. Inevitably, students run into issues when they test the robot they have built and programmed. Sometimes, students change the physical design of the robot to address an issue, and most times, they troubleshoot and revise the computer program they have written. Hence, design and problem solving are foundational activities when working in a robotics learning environment. Let us now consider the elements of design and problem solving as they relate to creativity.

9.4.1   Design

Design has been described by Svihla (2010) as "a ubiquitous activity that occurs in formal and everyday settings, commonly with a goal of making something to be used by someone else" (p. 246). In the case of robotics, students may focus on creating a number of different designs for different purposes. For example, they may design a robotic vehicle or device for use in a scientific investigation (e.g., data collection) (Mitnik, Nussbaum, and Recabarren Mitnik et al. 2009), or they may design a machine or vehicle capable of performing a certain type of work or solve specific challenges (Sullivan 2008; Sullivan 2011), or they may design a robotics system that models a biological system (Cuperman and Verner Cuperman and Verner 2013).

Schön (1983) has argued that framing a problem is at the heart of creativity in design. Framing a problem refers to seeing the problem from a specific perspective or with a specific lens. For example, the designer can choose to frame the problem from the perspective of a naïve or novice user of a product, and in this way, think about the basic functionality of a device. However, one may also choose to frame a problem from the perspective of the nature of the problem to be solved—what are the constituent elements of the problem and how can a design address each? Dorst and Cross (2001) have also argued that creativity in design unfolds as a process of problem framing. In their view, as people work on creating a design to address a specific problem their unfolding understanding of the problem informs the design of the solution. Dorst and Cross term this the co-evolution of the problem and the solution. This process of co-evolution is applicable to a robotics design situation for students. For example, I have found in my research that as students work in the robotics environment, they develop a better understanding of the problem to be solved (they frame or reframe the problem), and this has an influence on the design of the solution (Sullivan 2011).

Framing or reframing of a problem is foundational to design, and it can be thought of as an aspect of apposite design. Apposite design refers to a process in which a designer, or group of designers, arrives at a solution that has been hinted at, but not yet hit upon (hence, it's apposite nature). Such a solution may arise in the midst of a designing session when a designer has an illuminating thought in which the solution is brought into alignment with the constraints of the problem (co-evolution). For professional designers, this illuminating thought often has antecedents in sketches or discussions (Cross 1997), but arrives as a centralizing idea in the way that it bridges problem and solution. Akin and Lin (1995) have found that it is when designers are immersed in the triple process of sketching, examining, and thinking that they arrive at novel or illuminating design decisions, which, in many ways, rests on the knowledge of the designer. Given this, what are some ways that novice designers (e.g., students designing a robot) can develop the background knowledge and experience to help them develop a creative solution? One way to do this is to become familiar with design techniques. Cross (2004) defines four such techniques as follows:

- *Combination* in design is to take elements of existing designs and combine them to create a new design, for example, a headboard for a bed that has a built-in bookshelf.
- *Mutation* is when one takes one element of a designed artifact and changes it to create something new, for example, arm rests attached to the inside of a car door that have been expanded below to become compartments for small items.
- *Analogy* in design is best characterized by the design of the personal computer. Apple co-founder Steve Jobs conceived of the screen view on a computer as one that should be analogous to a person's actual desktop. In this way, the visual depiction and organizing principle of files and folders for the modern day computer was born.
- *First principles* approach refers to the idea of considering the domain that the designed item itself is a part of—Cross uses the example of sitting posture as first principles for the design of a chair. In using this approach, in order to design a chair, one should first consider the nature of the human body in a seated position—from there, one can develop a design. One problem with the first principles approach to design, as noted by Cross, is that, while it is possible to define first principles for a chair, it may be difficult to define them for other domains.

Each of these four design approaches requires synthesis of ideas. Owen (2007) argues that there are two types of creative activity in design: synthesis and discovery. An approach to design that relies more on discovery than synthesis is emergence. Cross (2004) discusses the idea of emergence, similar to apposite design, as the recognition of new properties within an existing design. These new properties can then be utilized as the basis for exploring new and different designs.

9.4.2   Social and Collaborative View of Design

While a designer may work alone, creativity in the process of design emerges from the social. This is so because as the designer works with ideas, sketches, or techniques, he is always drawing on the internally persuasive discourses that constitute his consciousness. Moreover, it is this consciousness, born of specific social and cultural voices that provide the individual designer with judgment and sensibilities that are then brought to bear on a design. We may think of this judgment and sensibility as the artistic aspects of design—irreducible to technical aspects only because of the very personal and unique origins of ideas. The social and collaborative view of design is also buttressed by the fact that a designer is always interacting with the reified ideas of other designers. For example, in the case of robotics, the student designers are interacting with the designed LEGO materials provided with the robotics construction kits, the design of these building materials cause them to fit together in specific ways, and these constraints suggest specific designs to builders. Within these material constraints, students find ways to realize their creative ideas.

In some ways, we can think of these constraints as an aspect of the "problem" of designing in the LEGO context. Co-evolution of a design, then, works both in terms of understanding the affordances and constraints of the materials, as well as the problem. Let us now consider the creative aspects of problem solving in greater detail.

9.4.3   Problem Solving

Creativity in problem solving may be manifested in multiple ways. Here we will consider three such modes that are relevant to robotics learning environments: problem finding, invented strategies, and idea generation.

9.4.4   Problem Finding

Similar to the notion of problem framing, problem finding has to do with identifying a problem, formulating an understanding of that problem, and clearly communicating ideas about the problem to others (Runco & Chand 1995). Csikszentmihalyi (1997) describes the process of problem finding as uncertain and emergent. However, there are things that individuals can do to enhance their problem-finding abilities. For example, Csikszentmihalyi argues that to develop a good formulation of a problem a person should think about it from as many perspectives as possible. In this way, students can create a number of possible paths to solutions to the problem. Moreover, students should not stay bound to one solution to a problem, rather they should evaluate the solution as it is unfolding. This is in the same spirit as Dorst and Cross's (2001) notion of the co-evolution of the problem discussed above. Changes or revisions to the solution should be made in relationship to how well the solution is working out.

In prior research, I found that students developed a creative solution to a problem through this process of co-evolution (Sullivan 2011). The key to the students' development of a creative idea was their evolving understanding of both the problem they were trying to solve and the functionality of the materials they were using to solve the problem. In this case, the students were asked to write a program that caused a light sensor to respond to specific environmental stimuli. In order to solve this problem, the students in the study first needed to transform their conceptual understanding of the light sensor as a measurement device to a computational device. They also needed to reframe the problem accordingly; rather than simply measuring the amount of light that was reflecting off of one surface, the students needed to measure the light reflected off of both the approach surface and the target surface, and they needed to use the differential between those two readings to program the robot to respond to the smaller (less reflected light) reading. In this case, the students discovered that their target surface was not as dark as required to solve the problem. This discovery is what Koschmann and Zemel (2009) would call an *occasioned production*. It is a discovery that the students did not know they needed to make prior to the moment they made it. Once the students had made this discovery and had developed a greater understanding of how the light sensor worked, they were able to repurpose materials provided to them in the learning environment to create a darker target surface and thereby solve the robotics challenge. In this case, the repurposing of materials constituted a creative solution to the problem.

Research on problem finding indicates that the type of problem to be solved will have an effect on the creativity of the solution. For example, ill-structured or open problems require greater problem-finding efforts, and this correlates with more creative solutions, whereas well-structured or closed problems call for more reproductive or algorithmic solutions (Jay & Perkins 1997; Mumford, Blair, & Marcy Mumford et al. 2006; Ward, Finke, & Smith 1995; Weisberg 1986). According to Jonassen (2000), an ill-structured problem has the following characteristics:

- Not all of the problem elements are known.
- There are multiple solutions and multiple-solution paths.
- There are multiple criteria for evaluating the solution.
- Solutions may require personal judgment on the part of the solver.

Robotics challenges may be considered ill-structured in that there are always multiple solutions and multiple-solution paths. This is so as equally successful solutions to a robotics challenge may include differently constructed robotic devices and differing computer programs. Because of this, it is possible that students solving robotics problems will be more likely to develop their own solutions.

Amabile (2012) discusses this as heuristic versus algorithmic problem solving. When tackling an open-ended problem, people have the chance to develop their own heuristic approach to the solution; this is often a creative endeavor in itself, whereas simply applying an existing, step-by-step algorithm to solve a closed problem does not require much creativity. Let us now consider in greater detail the role of invented strategies in robotics.

9.4.5 Invented Strategies

While many students begin their robotics problem-solving efforts using a trial-and-error method (Barak & Zadok 2009; Castledine & Chalmers 2011; Gaudiello & Zibetti 2013; Sullivan & Lin 2012; Williams, Ma, Prejean, Ford, & Lai 2008), over time, they move beyond that method and begin to develop more sophisticated approaches to problem solving. Barak and Zadok identified two approaches that they call heuristic searches—defined as approaches that leverage the knowledge students have built about the problem to help them solve the problem. The first type of heuristic search, called the proximity method, involves forward and backward reasoning toward the goal of gradually arriving at a solution. The second approach involves planning that includes modeling and reasoning through analogy or abstraction.

In our work, we also identified students using a modeling strategy to reason about the problem (Sullivan & Lin 2012). Students used either the robotics materials or their own bodies to simulate the desired movement of the robot. For example, we have repeatedly observed students hold a robotic device in their hands and move it along a surface in the manner they would like to program it to run. In this way, students think through the discreet movements that the robot will execute and they consider how they must program the robot to do so. Oftentimes, students will go back and forth between moving the robot by hand and writing the program. This strategy of simulating the movement of the robot may also take place with a student's hand standing in for the robot. In other words, we have observed students moving their hands in the manner that the robot should be programmed to move—this activity serves the same intellectual purpose as physically moving the robotic device.

The invented strategy of using one's own body to model/simulate the movement of the robot is an aspect of embodied cognition. Embodied cognition is defined by Weiskopf (2010) as the view that "cognitive capacities are shaped and structured by the bodily capacities of a creature, including the sensorimotor capacities that make possible its basic interactions with the world" (p. 295). The theoretical basis for embodied cognition is provided by Barsalou's (2003) situated-simulation theory. In this formulation, cognition is not characterized by, what Barsalou has termed, amodal semantic knowledge (memorized texts), but, rather, by modal recall (simulation) of experiences and actions on a sensorimotor level. Barsalou argues that:

When the conceptual system represents an object's visual properties, it uses representations in the visual system; when it represents the actions performed on an object, it uses motor representations. The claim is not that modal reenactments constitute the sole form of conceptual representation...The claim is simply that modal reenactments are an important and widely utilized form of representation (p. 521).

Embodied cognition is enabled by robotics devices, and many students will intuitively use their own bodies to reason about how to program the robotic device. In this way, the invented strategy is not only an instance of creative problem solving, but also it may lead to increased learning because it activates real-world knowledge and improves memory and recall through physical action (McNeil and Jarvin 2007).

9.4.6 Idea Generation

A third mode of creative activity related to problem solving is idea generation. As discussed at the beginning of this chapter, the sociocultural approach to understanding creativity focuses, in part, on the discursive interactions among people. Rojas-Drummond, Albarrán, and Littleton (2008) in their research on collaborative creativity in the language classroom have developed a good explanation of how idea generation works in a group:

Among the common acts present in all the [discourse] data were: joint planning; taking turns; asking for and providing opinions; sharing, chaining and integrating of ideas; arguing their points of view; negotiating and coordinating perspectives; adding, revising, reformulating and elaborating on the information under discussion and seeking of agreements. These data, taken together, suggest that the children engaged in diverse processes of "co-construction" of meaning and knowledge to achieve their goals (p. 186).

The evolution of a creative idea, then, rests in part on an interactive process in which students draw on their own prior knowledge (internally persuasive discourses) and experience to generate ideas and contribute them to the group. These ideas may then be acted upon by others in a number of ways including evaluation, elaboration, clarification, or refutation. Moreover, these ideas may then be considered in light of the other voices that are present in the classroom that contribute to the development of knowledge about the problem. Indeed, in my research, I have found that a number of voices in a classroom can contribute to the development of students' creative ideas including the voices of collaborative group members, voices of other students in the classroom, the voice of the classroom teacher, the reified voice of the curriculum developers, and the reified voices of the technology designers (Sullivan 2011). It is the interanimation, the juxtaposition of these voices with one another that may lead the students to the development of a creative solution.

Having considered the creative aspects of the twin robotics activities of design and problem solving, let us now consider how to support student creativity with robotics in terms of curriculum and pedagogy.

9.5 Curricular and Pedagogical Approaches to Support Creativity

9.5.1 Robotics Curricula

As argued elsewhere (Sullivan & Heffernan 2016), teachers may use robotics to teach about robotics (first-order uses), or to teach about other subjects (second order uses). In terms of first-order uses, curricula may focus on issues related to the engineering design of a robotics system, including how to create a working gear system, building stable and functional structures, and the steps one should take to create and troubleshoot the device (the engineering design process) (Heffernan 2013). First-order uses will also focus on teaching about programming the robot and may include instruction on sequencing (Kazakoff & Bers 2012; Kazakoff, Sullivan, & Bers 2013), conditional reasoning with sensors (Slangen, van Keulen, & Gravemeijer 2011; Sullivan & Lin 2012), and computer science concepts such as input/output/process, control structures, iteration, and parallel programming (Nugent, Barker, Grandgenett, & Adamchuk 2010; Sullivan & Lin 2012).

Meanwhile, second-order uses of robotics in the curriculum generally focus on modeling or simulating systems or phenomena in the natural sciences. For example, one may use robotics to model biological systems (Cuperman & Verner 2013) or to teach about forces and motion in a physics context (Mitnik, Nussbaum, & Recabarren 2009; Williams, Ma, Prejean, Ford, & Lai 2008). Moreover, robotics may be used to reinforce students understanding of systems in the context of science literacy (Sullivan 2008).

9.5.2 Curricular Enactment and Pedagogical Approach

Whether a teacher chooses to use robotics to teach about robotics or some other topic, certain elements of the curricular enactment and pedagogical approach are key to supporting student creativity. These key elements are (1) the nature of the problem to be solved; (2) student choice; (3) a non-evaluative stance; (4) the nature of student interactions; and (5) modeling or allowing playful inquiry.

Nature of the problem to be solved. As noted earlier in this chapter, problems may be well-structured or ill-structured (Jonassen 2000). Many times, in formal PK-12 school settings, children and youth are asked to solve well-structured problems. Well-structured problems have the following characteristics:

- All elements of the problem are readily apparent
- Problem solution requires the application of a limited number of rules or principles
- Problem solutions are knowable and comprehensible, and all problem states are probabilistic

While it is possible to develop a meaningful and rich, well-structured robotics problem for students to solve (Sullivan 2005), ill-structured problems are the better choice for enabling creativity. This is so as ill-structured problems support many paths to solution and many correct answers. As noted earlier in this chapter, in a design context, creativity is an aspect of the co-evolution of one's understanding of the problem and the solution. If students are provided with a problem where all elements of the problem are readily apparent, they will not need to evolve their understanding of the problem. Therefore, providing a problem that is robust enough to support many paths to solution will expand the learning space and the learning opportunities for students, thereby improving the chances that students well engage in creative design.

Student choice. To support student creativity, it is critical that students be given some choice in the robotics project they will undertake. Providing students opportunities to choose how they want to work with robotics allows them to follow their intrinsic interests, which fosters creativity (Amabile 1983). Providing choice does not preclude teacher facilitation or scaffolding; indeed, it is very important that the teacher works closely with students as they evolve the project they want to accomplish. A project-based approach to robotics centers on student-driven questions about robotics. The role of the teacher is to consider the knowledge, experiences, and interests of the students and build on these through robotics projects that are open-ended, diverse, and challenging. A project-based approach that allows students to explore robotics over time will support student learning, interest, and creativity.

Non-evaluative stance. Another important pedagogical aspect of supporting student creativity is to take a non-evaluative stance. While such a stance is anti-thetical to the measurement focus of regular classwork, it may be possible to run a robotics project at a time of year when evaluation can be set aside and interest-based learning can take center stage. In the USA, such a time comes in late spring when students have finished taking all of the standardized tests expected of them. During these last six weeks of the year, many teachers are able to undertake more interesting activities with their students. This is a good time to devise a robotics project that will reinforce the learning that students have already done during the year (as in a second order approach), or to simply introduce them to the field of robotics (as in a first-order approach). The goal is to allow students to explore, learn, and create without the tension of evaluation and expectation. Hennessey (1995) has shown that student creativity is enhanced when they have a chance to pursue their own intrinsic interest in a non-evaluative, open inquiry. Notably, both of these elements are very important, students must be allowed to pursue their own interests, and their efforts need not be evaluated—when these two classroom conditions are met, student's creativity will be realized.

Nature of student interactions. In addition to creating a classroom environment that allows for student choice in a non-evaluative format, to enable and support creativity with robotics, students must be allowed to work together. Students may be supported to work with one another in collaborative groups, and/or they may be allowed to engage in what Zhang, Scardamalia, Reeve, and Messina (2006) have called opportunistic collaboration. This form of collaboration supports spontaneous collaborative interactions among any students in the classroom. Students may choose to work with one another to achieve a specific goal, or one or more students may choose to observe the work of other students in order to learn and develop a better understanding of robotics. The goal is to create an open classroom where students can work with and learn from one another. The operative elements of such a classroom are freedom of movement for students. In other words, while they are working on their robotics projects, students should be allowed to freely move about the room and to speak with any student they choose. While there is potential in this situation for students to veer off course and not tend to the robotics project, there is also great potential for students to engage in deep inquiry and learning. Skillful teachers will find ways to support students in staying on task with their chosen robotics projects.

Modeling or allowing playful inquiry. Teachers should not be alarmed if the robotics learning activities that students engage in resemble play. As has been noted earlier in this chapter, play is a meaningful mode of inquiry in the field of robotics, and I would argue, whenever one wishes to improve or enable creativity. The teacher may also engage in play with the students to aid their learning. For example, in my prior research (Sullivan 2011), I have witnessed the power of the teacher's playful inquiry to both motivate students to work with robotics and also in causing them to develop working analogies to more fully explore and build knowledge about the robot as well as the task at hand. In this particular example, the teacher playfully triggered a program on the robot by passing his shod foot below the light sensor affixed to the front of a robotic device. In playing with the robot, the teacher spurred the children to interact in a similar fashion, and in this way, they developed an understanding of the light sensor as functioning like a magnet, which, as noted earlier, is an accurate analogy when the robotic device is running a line following program (as it was programmed to do in this case).

9.6   Conclusion

In this chapter, I have made the case for the creative nature of learning with robotics. Robotics is an interdisciplinary STEM activity, consisting of both engineering design and computer programming content and concepts. Indeed, at the heart of robotics learning is design and problem solving, two activities that have a strong potential to enable student creativity. This creativity may begin through play —as noted above, students may first spend time playing with the robotics devices in order to learn about them or to orient themselves in relation to robotics.

Teachers must keep in mind that to actually enact a robotics curriculum that engenders play and creativity, the curricular and pedagogical approaches must include ill-structured problems that students have chosen to tackle due to their own interests. Moreover, the classroom environment should provide students opportunities to walk around and collaborate freely with other students. Teachers should withhold evaluation of students' ideas, and rather encourage and support them in their work; finally, teachers should allow students time to play with the robotic devices; and, if so inclined, the teacher may choose to model such playful behavior. If such a playful, open, nonjudgmental, and collaborative environment is created, the potential for student creativity with robotics will be high.

References

Akin, Ö., & Lin, C. (1995). Design protocol data and novel design decisions. *Design Studies, 16*(2), 211–236.

Amabile, T. M. (1983). The social psychology of creativity: A componential conceptualization. *Journal of Personality and Social Psychology, 45*(2), 357–376.

Amabile, T. (2012). *Componential theory of creativity*. Boston, MA: Harvard Business School.

American Association of University Women. (2010). *Why so few? Women in Science Technology Engineering and Mathematics*. Washington, DC: Author.

Bakhtin, M. M. (1981). *The dialogic imagination* (C. Emerson & M. Holquist, Trans.). Austin, TX: University of Texas Press.

Bakhtin, M. M. (1986). The problem of speech genres. In C. Emerson & M. Holquist (Eds.), *Speech genres and other late essays* (V.W. McGee, Trans. pp. 60–106). Austin, TX: University of Texas Press.

Barak, M., & Zadok, Y. (2009). Robotics projects and learning concepts in science, technology and problem solving. *International Journal of Technology and Design Education, 19*(3), 289–307.

Barsalou, L. (2003). Situated simulation in the human conceptual system. *Language and Cognitive Processes, 18*(5/6), 513–562.

Castledine, A. R., & Chalmers, C. (2011). LEGO Robotics: An authentic problem-solving tool? *Design and Technology Education, 16*(3), 19–27.

Cross, N. (1997). Creativity in design: Analyzing and modeling the creative leap. *Leonardo, 30*(4), 311–317.

Cross, N. (2004). Expertise in design: An overview. *Design Studies, 25*(5), 427–441.

Csikszentmihalyi, M. (1997). *Finding flow: The psychology of engagement with everyday life*. New York, NY: Basic Books.

Cuperman, D., & Verner, I. M. (2013). Learning through creating robotic models of biological systems. *International Journal of Technology and Design Education, 23*, 849–866.

Dorst, K., & Cross, N. (2001). Creativity in the design process: Co-evolution of problem–solution. *Design Studies, 22*(5), 425–437.

Gaudiello, I., & Zibetti, E. (2013). Using control heuristics as a means to explore the educational potential of robotics kits. *Themes in Science & Technology Education, 6*(1), 15–28.

Heffernan, J. (2013). *Elementary engineering: Sustaining the natural engineering instincts of children*. Charlestown, SC: Printed by CreateSpace.

Hennessey, B. A. (1995). Social, environmental, and developmental issues and creativity. *Educational Psychology Review, 7*, 163–183.

Jay, E. S., & Perkins, D. N. (1997). Problem finding: The search for mechanism. In M. A. Runco (Ed.), *The creativity research handbook* (Vol. one, pp. 257–294). Cresskill, NJ: Hampton Press.

Jonassen, D. H. (2000). Toward a design theory of problem solving. *Educational Technology Research and Development, 48*(4), 63–85.

Kazakoff, E., & Bers, M. U. (2012). Programming in a robotics context in the kindergarten classroom: the impact on sequencing skills. *Journal of Educational and Hypermedia, 21*(4), 371–391.

Kazakoff, E. R., Sullivan, A., & Bers, M. U. (2013). The effect of a classroom-based intensive robotics and programming workshop on sequencing ability in early childhood. *Early Childhood Education Journal, 41*(4), 245–255.

Koschmann, T., & Zemel, A. (2009). Optical pulsars and black arrows: Discoveries as occasioned productions. *Journal of the Learning Sciences, 18*(2), 200–246.

Kristeva, J. (1986). Word, dialogue and novel. In T. Moi (Ed.), *The Kristeva reader* (pp. 34–61). New York, NY: Columbia University Press.

McNeil, N., & Jarvin, L. (2007). When theories don't add up: Disentangling the manipulatives debate. *Theory into Practice, 46*(4), 309–316.

Mitnik, R., Recabarren, M., Nussbaum, M., & Soto, A. (2009). Collaborative robotic instruction: A graph teaching experience. *Computers and Education, 53,* 330–342.

Mumford, M. D., Blair, C. S., & Marcy, R. T. (2006). Alternative knowledge structures in creative thought: Schema, associations and cases. In J. C. Kaufman & J. Baer (Eds.), *Creativity and reason in cognitive development* (pp. 117–136). New York, NY: Cambridge University Press.

Norman, D. A. (2002). *The design of everyday things.* New York, NY: Basic Books.

Nugent, G., Barker, B., Grandgenett, N., & Adamchuk, V. (2010). Impact of robotics and geospatial technology interventions on youth STEM learning and attitudes. *Journal of Research on Technology in Education, 42*(4), 391–408.

Owen, C. (2007). Design thinking: Notes on its nature and use. *Design Research Quarterly, 2*(1), 16–27.

Resnick, M. (2006). Computer as paintbrush: Technology, play, and the creative society. In D. Singer, R. Golikoff, & K. Hirsh-Pasek (Eds.), *Play = learning: How play motivates and enhances children's cognitive and social-emotional growth.* Oxford: Oxford University Press.

Resnick, M., Maloney, J., Monroy-Hernandez, A., Rusk, N., Eastmond, E., Brennan, K., et al. (2009). Scratch: Programming for all [Electronic version]. *Communications of the ACM, 52* (11), 60–67.

Resnick, M., & Martin, F. (1991). LEGO/logo: Learning through and about design. In I. Harel & S. Papert (Eds.), *Constructionism* (pp. 183–192). Norwood, NJ: Ablex.

Rojas-Drummond, S. M., Albarrán, C. D., & Littleton, K. S. (2008). Collaboration, creativity and the co-construction of oral and written texts. *Thinking Skills and Creativity, 3*(3), 177–191.

Runco, M. A., & Chand, I. (1995). Cognition and creativity. *Educational Psychology Review, 7*(3), 243–267.

Rusk, N., Resnick, M., Berg, R., & Pezalla-Granlund, M. (2008). New pathways into robotics: Strategies for broadening participation. *Journal of Science Education and Technology, 17*(1), 59–69.

Schön, D. A. (1983). *The reflective practitioner: How professionals think in action.* New York, NY: Basic books.

Slangen, L., van Keulen, H., & Gravemeijer, K. (2011). What pupils can learn from working with robotic direct manipulation environments. *International Journal of Technology and Design Education, 21*(4), 449–469.

Sullivan, F. R. (2005). *The ideal science student and problem solving.* Unpublished doctoral dissertation, Teachers College, Columbia University, New York New York.

Sullivan, F. R. (2008). Robotics and science literacy: Thinking skills, science process skills, and systems understanding. *Journal of Research in Science Teaching, 45*(3), 373–394.

Sullivan, F. R. (2011). Serious and playful inquiry: Epistemological aspects of collaborative creativity. *Journal of Educational Technology and Society, 14*(1), 55–65.

Sullivan, F. R., & Heffernan, J. (2016). Robotic construction kits as computational manipulatives for learning in the STEM disciplines. *Journal of Research in Technology Education, 49*(2), 105–128. doi:10.1080/15391523.2016.1146563.

Sullivan, F. R., & Lin, X. D. (2012). The ideal science student survey: Exploring the relationship of students' perceptions to their problem solving activity in a robotics context. *Journal of Interactive Learning Research, 23*(3), 273–308.

Sullivan, F. R., & Wilson, N. C. (2015). Playful talk: Negotiating opportunities to learn in collaborative groups. *Journal of the Learning Sciences, 24*(1), 5–52.

Svihla, V. (2010). Collaboration as a dimension of design innovation. *Journal of CoDesign: International Journal of CoCreation in Design and the Arts, 6*(4), 245–262.

Ward, T. B., Finke, R. A., & Smith, S. M. (1995). *Creativity and the mind: Discovering the genius within.* New York, NY: Plenum Press.

Weisberg, R. (1986). *Creativity: Genius and other myths.* New York, NY: W.H. Freeman and Company.

Weiskopf, D. A. (2010). Embodied cognition and linguistic comprehension. *Studies in History and Philosophy of Science Part A, 41*(3), 294–304. doi:10.1016/j.shpsa.2010.07.005.

Williams, D. C., Ma, Y., Prejean, L., Ford, M. J., & Lai, G. (2008). Acquisition of physics content knowledge and scientific inquiry skills in a robotics summer camp. *Research on Technology in Education, 40*(2), 201–216.

Zhang, J., Scardamalia, M., Reeve, R., & Messina, R. (2006). *Collective cognitive responsibility in knowledge building communities.* American Educational Research Association Annual Meeting, San Francisco, CA.

Chapter 10
Dancing, Drawing, and Dramatic Robots: Integrating Robotics and the Arts to Teach Foundational STEAM Concepts to Young Children

Amanda Sullivan, Amanda Strawhacker and Marina Umaschi Bers

Abstract In recent years, there has been an increasing national focus on the importance of Science, Technology, Engineering, and Math (STEM) education for young children beginning in kindergarten. This chapter explores the newest acronym, "STEAM," which integrates the Arts with STEM education. While many assume the "A" in STEAM refers only to the fine arts, the full potential of STEAM goes beyond aesthetics to include language arts, culture, history, and the humanities. The emerging domain of robotics offers playful strategies for engaging young children with the technology and engineering components of STEM. Additionally, when implemented thoughtfully, robotics is a creative medium with the power to engage young children in the arts and humanities. KIBO is a newly developed robotics construction set specifically designed for children ages 4–7 years to learn foundational engineering and programming content in a hands-on, open-ended way —no screen-time required! This chapter presents vignettes of three interdisciplinary robotics curricular units that utilize the KIBO Robotics Kit: (1) *Dances from Around the World*, (2) *Art-Making Robots*, and (3) *Superhero Bots*. It highlights strategies for taking a child-focused approach to robotics education by drawing on student interest in music, visual arts, and literature when exploring foundational technological ideas.

Keywords Robotics · Early childhood · Humanities · Arts · STEAM

A. Sullivan (✉) · A. Strawhacker · M.U. Bers
Tufts University, Medford, MA, USA
e-mail: amanda.sullivan@tufts.edu

10.1   Introduction

The difference between science and the arts is not that they are different sides of the same
coin [...] or even different parts of the same continuum, but rather, they are manifestations
of the same thing. The arts and sciences are avatars of human creativity.

—Mae Jemison, doctor, dancer, and first African-American woman in space

Science, technology, engineering, and mathematics (STEM) education has been of
growing importance to educators and researchers working with young children. In
recent years, there has been a particular focus on addressing the "T" of technology and
"E" of engineering in early childhood education through robotics and computer programming applications. This is partially due to federal education programs and private
initiatives making computer science and technological literacy a priority for young
children in a growing number of countries worldwide (e.g., U.S. Department of
Education 2010; UK Department of Education 2013). However, as technology has
grown in prevalence in schools and at home, some researchers and educators have also
expressed concern that excessive usage of computers and digital technologies may
actually stifle children's learning and creativity through passive consumption of media
(Cordes and Miller 2000; Oppenheimer 2003). In order to address these concerns, there
is a growing body of work on how technology can be used to foster positive behaviors
and engage children as *creators* rather than *consumers* of digital content (Bers 2012;
Resnick 2006). One way to do this is by explicitly integrating the arts, self-expression,
and identity exploration with traditional technology and engineering curricula for
young children.

Over the past few years, there has been growing enthusiasm about integrating the
arts with STEM in early childhood settings. This trend is clear in school curricula,
new educational initiatives, and even in popular children's media. In the 43rd
season of *Sesame Street* (which was aired from 2012 to 2013), the television show
continued its introduction of STEM education with the addition of arts, introducing
viewers to the acronym of STEAM (Science, Technology, Engineering, *Arts*, and
Mathematics) for the first time on the show (Maeda 2012). This milestone for the
STEAM movement was achieved surprisingly quickly, as it was only one year prior
that Elmo was interviewed on CNN about the importance of STEM (Maeda 2012).

The STEAM movement was originally spearheaded by the Rhode Island School
of Design (RISD) but is now widely adopted by schools, businesses, and individuals (STEM to STEAM 2016). In some ways, it is surprising that the STEM
education movement has not always included an integration of the arts. Historically,
there have been a countless brilliant innovators, such as Leonardo Da Vinci and
Frank Lloyd Wright, who have woven together the fields of art and science
seamlessly in their work. Many modern innovations have also resulted from an
integration of the arts with STEM. For example, the computer chips that run almost
all of our digital devices are made using a combination of three classic artistic
inventions: etching, silk screen printing, and photolithography (Root-Bernstein
2011). Today, it is estimated that Nobel laureates in the sciences are seventeen

times likelier than the average scientist to be a painter, twelve times as likely to be a poet, and four times as likely to be a musician (Pomeroy 2012). This is likely because the arts, such as science, technology, engineering, and mathematics, are rooted in a similar mindset of curiosity and creativity.

It is clear that the arts have a place in the exploration of technology and the sciences, but *how* educators bring these fields together effectively can be a challenging question, particularly during the early childhood years (ages 4–7 years) when teachers are already juggling a large load of foundational content that needs to be covered. This chapter describes how robotics can be used as a creative medium for young children to playfully explore STEAM content in a developmentally appropriate way. It presents three illustrative vignettes that highlight different interpretations of the "A" in STEAM, all using the newly developed KIBO robotics kit developed by the DevTech Research Group at Tufts University and KinderLab Robotics (Sullivan et al. 2015). The goal of this work is to provide readers with examples of how robotic tools like KIBO can facilitate STEAM learning in a natural way that can easily tie in with content educators are already teaching. Additionally, it demonstrates examples of how STEAM curricula can be implemented in different learning environments such as formal classrooms, camps, and extracurricular clubs.

10.2   Literature Review

10.2.1   Moving from STEM to STEAM in Early Childhood

Historically, early childhood education has focused on building foundational numeracy skills and an understanding of the natural sciences when it came to STEM (Science, Technology, Engineering, Mathematics) education for young children (Bers 2008; Bers et al. 2013). In the growing national-level discussion around STEM, the question of how to teach technology and engineering has become more pressing (UK Department of Education 2013; US Department of Education 2010). New education policy changes, commercial products, and non-profit organizations are promoting a message that highlights the benefits of computational thinking, digital citizenship, and technological literacy (Bers 2012; Hobbs 2010; Hollandsworth et al. 2011; White House 2011; Wing 2006). However, this increased focus on children's usage of computers and digital technology has also sparked some concern that children's natural play and creativity may be stifled by these tools (Cordes and Miller 2000; Oppenheimer 2003). Adding the arts to STEM-based subjects, such as computer programming and engineering, may enhance student learning by infusing opportunities for creativity and innovation (Robelen 2011).

This concept of promoting creativity and expression through technology is articulated in a newer acronym called "STEAM" (Science, Technology, Engineering, *Arts*, Mathematics) that is growing in popularity (Yakman 2008). The "A" of STEAM can represent more than just the visual arts, but rather a broad spectrum of the humanities including the liberal arts, language arts, social studies, music, culture, and more. For example, Maguth (2012) proposed that social studies content should also be integrated into a STEM-focused curriculum in order to promote the development of well-rounded citizens prepared for voting on ethical and social issues related to STEM. New technologies, such as programmable robotics kits described in the following section, offer innovative ways to integrate the arts with traditional technology and engineering content.

10.2.2 Robotics in Early Childhood Education

Early childhood is an important time to explore the arts and play, as children need hands-on experiences to construct their own learning (Papert 1980). Although robotics and programming can seem rigid, there is much room in these fields for creativity and innovation (Resnick 2006). Digital media, when designed within developmentally appropriate guidelines, can afford children the same opportunities for exploration and construction that traditional learning tools offer (Bers 2008). In research trials with simple robotics and programming languages, children as young as 4 years old demonstrated understanding of foundational engineering and robotics content, (Bers et al. 2002; Sullivan et al. 2013; Sullivan and Bers 2015; Cejka et al. 2006; Perlman 1976; Wyeth 2008). In addition to mastering this new content, programming interventions have been shown to have positive benefits for children's developing numeracy, literacy, and visual memory, and can even prompt collaboration and teamwork (Clements 1999; Lee et al. 2013).

New developmentally appropriate technological software and robotic kits have evolved in the tradition of educational manipulatives such as Froebel's "gifts," Montessori materials, and Nicholson's loose parts, and like their predecessors, these tools allow children to explore their understanding of shape and number, spatial relations, and proportion (Kuh 2014; Nicholson 1972; Resnick et al. 1998; Brosterman 1997). However, cognitive development is not the only area of growth for young children, and "screen time" is a serious concern for learners at this age (American Academy of Pediatrics 2003). Robotics curricula are now being developed to address children's need to move, dance, and push their physical boundaries. For example, when constructing a robot with many parts, children may exercise fine motor skills, and when observing a robot's movements, children are compelled to move and dance along, developing their hand-eye coordination and gross-motor activity (Resnick et al. 1998).

10.2.3   The KIBO Robotics Kit

KIBO is a robotics kit designed specifically to playfully introduce young children (ages 4–7 years) to foundational engineering, programming, and computational thinking concepts through tangible "screen-free" activities (Bers 2017; Sullivan et al. 2015). KIBO was created based on research by the Developmental Technologies Research Group at Tufts University and made commercially available through funding from the National Science Foundation (NSF) and a successful Kickstarter campaign (Sullivan et al. 2015). KIBO is unique as compared to its counterparts on the commercial market because it engages children with both building with robotic parts (KIBO's hardware) and programming KIBO to move with tangible programming blocks (KIBO's software). KIBO is designed based on years of child development research at Tufts University and is intended explicitly to meet the developmental needs of young children (e.g., Sullivan and Bers 2015; Sullivan et al. 2015; Kazakoff and Bers 2012). The kit contains easy-to-connect construction materials including wheels, motors, light output, a variety of sensors, and wooden art platforms (see Fig. 10.1).

KIBO is programmed to move using a tangible programming language that consists of eighteen interlocking wooden programming blocks and 12 parameters (see Fig. 10.2). With just eighteen blocks, children are able to master increasingly complex computational thinking concepts such as repeat loops, conditional statements, and nesting statements (Bers 2017). These wooden blocks resemble familiar early childhood manipulatives such as alphabet blocks and contain no embedded electronics or digital components. Instead, KIBO's main body has an embedded scanner that scans the barcodes on the programming blocks in order to "read" the program the child has

Fig. 10.1 The KIBO robot with sensors, light output, and sample block program

Fig. 10.2 KIBO's tangible programming language

created. Once the program has been scanned, it is saved on the robot instantly. No interaction with a computer, tablet, or other screen-based software is required to learn programming with KIBO. This tangible approach to computer programming is developmentally appropriate for young children and is aligned with the American Academy of Pediatrics' recommendation that young children have a limited amount of screen time per day (American Academy of Pediatrics 2003).

10.2.4 Exploring STEAM Through KIBO

The KIBO robot is well suited to exploring a variety of STEAM concepts. From a technology and engineering perspective, children can use the kit to learn about basic electronic elements they encounter everyday but may not understand, such as wires, batteries, computer chips, motors, sensors, and light bulbs. Additionally, children can explore foundational programming concepts such as sequencing, repeat loops, and conditional statements in order to make their robot move.

Along with these robotic and programming components, the KIBO kit also contains art platforms that can be used for children to personalize their projects with craft materials in order to foster STEAM integration. Unlike KIBO, many robots for young children come out of the box already decorated to look like a toy or creature. For example, the Beebot (a programmable floor robot developed for preschoolers) is designed to look like a bumblebee. The Wonder Workshop robots Bo and Yana (robots programmed through an ipad application) are round and blue with large eye in the center, resembling a friendly creature. KIBO on the other hand does not look like anything until the child places his or her imagination on it. It does not have a face and is constructed of neutral colors and wooden materials, in the style of Reggio-Emilia child manipulative design (Strong-Wilson and Ellis 2007; Kuh 2014). Like an unsculpted wad of play-dough, KIBO looks and feels differently each time the child uses it, which makes the robot ideal for an integration of the arts (see Figs. 10.3 and 10.4). The following section will provide three vignettes of the KIBO robot as it has been used in different types of playful learning settings to explore STEAM content.

10.2.5   Designing STEAM Curricula with KIBO

It can be challenging to design STEAM activities for young children that not only promote technological content, but also foster elements of personal and interpersonal skills that are critical to early childhood development. Bers' (2012) Positive Technological Development Framework (PTD) provides a framework to guide educators in the creation of effective pedagogy and classroom practices. The PTD framework is rooted in the field of applied child development and is based on the Positive Youth Development framework created by Lerner et al. (2005) and Constructionist theory (Papert, 1980). PTD focuses on enhancing positive skills and behaviors in children by describing "6 Cs" that the digital world offers to promote healthy development in our youth: communication, collaboration, community building, content creation, creativity, and choices of conduct. Each of the vignettes in the following section feature curriculum that was developed with these 6 C's in

Fig. 10.3 Sample KIBO construction

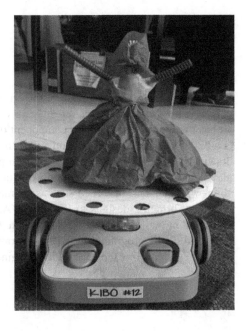

Fig. 10.4 Sample KIBO
kinetic sculpture

mind. They provide examples of children not only engaging with content creation through robotics and the arts, but also communicating with one another, sharing their work with a larger community, and creatively following their passions.

10.3 Vignette 1: Art-Bots

10.3.1 Curriculum Overview

Children can gain their first taste of the art world by exploring, manipulating, and playing with tactile and visual art materials. From finger-painting to sculting clay, art has traditionally been a core component of early childhood education (Althouse et al. 2003). In the early childhood years, the visual arts are typically composed of drawing, painting, arts and crafts, and sculpting with materials such as clay and play-dough (2003). This exploration of drawing, painting, and crafting naturally aligns with elements of STEM such as geometry, engineering sturdy structures, and iteratively bringing designs to life. In the Art-Bots curriculum, the visual arts are integrated in two specific ways: (1) through the design and decoration of the KIBO robot and (2) by programming KIBO to draw and paint on paper. This curriculum focuses explicitly on the technology, engineering, math, and art components of STEAM.

10.3.2 Educational Environment

The art-making robots activity can apply to a variety of educational settings including classrooms, clubs, and at home. In this example, Art-Bots were explored in an extracurricular Saturday extracurricular club called the "Saturday STEM Series." Over the course of five weeks, the Saturday STEM club met once a week for a six-hour period to explore different concepts such as the engineering design process, animations and storytelling, robotics, programming, and more. Unlike formal school classrooms, this STEM series was open to a mixed-age group of children in Kindergarten through second grade. Therefore, the series was typically made up of a diverse group of approximately 10 children including boys and girls from both public and private schools ranging from 5 to 7 years old. This mixed-age learning setting more closely resembled a Montessori classroom or an informal play environment than a traditional classroom.

10.3.3 STEAM Concepts Explored

This STEM-Saturday session was scheduled as a culmination to prior sessions that introduced foundational KIBO robotics and programming concepts. Children were already familiar with this robotic kit and basic concepts of coding and engineering. For this reason, this activity was not an introductory robotics exploration, but more like creative free play using a familiar technological tool.

10.3.4 The Engineering Design Process

The Art-Bots curriculum introduced young children to the engineering design process by giving children the challenge of building a 3-dimensional structure on top of KIBO's art platforms. This structure could be motorized or static, but needed to be secure enough that when KIBO executed a program the structure stayed securely connected to the robot. This encouraged children to focus on the steps of the engineering design process. In particular, children practiced testing and improving their robot by trying out various types of materials, testing out different programs, and experimenting with ways to attach the structures to their robot.

The engineering design process provides children with a cyclical way to satisfy their budding curiosity about the world by asking questions or posing problems that they are personally interested in investigating (see Fig. 10.5). Early childhood is an ideal time to begin teaching engineering concepts because children are naturally inquisitive about the world around them and are motivated to explore, build, and discover answers to their big questions. The engineering design process refers to the iterative process engineers use to design an artifact in order to meet a need (Bers

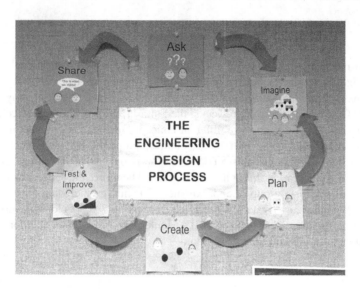

Fig. 10.5 Engineering design process classroom poster

2010). Although it is a fluid process, its steps typically include some variation of the following: identifying a problem, looking for ideas for solutions and choosing one, developing a prototype, testing, improving, and sharing solutions with others (MA DOE 2006). The curriculum adapted this definition for use with children ages 5–7 years. The STEM-Saturday instructors paid particular attention to the steps of testing and improving, which require problem solving and perseverance, critical skills for young children's social and emotional development.

Children were given access to a range of materials including paper, play-dough, legos, tissue paper, cardboard, and other recycled and craft materials. They were given open-ended prompting about the type of structure they could create; therefore, the challenge was interpreted in a variety of ways. While some children created little kinetic sculptures of things such as people and animals (see Figs. 10.6 and 10.7), other children created more abstract decorations (see Figs. 10.8 and 10.9).

10.3.5 Geometry

Once children felt comfortable with artistic design and decorating their robots, the unit continued with an exploration of creating lines and shapes through movement. Children brainstormed the different two-dimensional shapes they were familiar with and practiced drawing squares, circles, trapezoids, diamonds, and more on paper. Then, they were presented with this question: Can KIBO draw any of these shapes? As a group, the class made hypotheses as to which shapes KIBO could or could not

Fig. 10.8 Abstract KIBO decoration

Fig. 10.9 Mixed-media KIBO decoration

draw with its current programming language. This led to a group dialogue on KIBOs ability to draw a circle, with some children arguing that KIBO's straight-line movements could not allow for a curved shape.

To test these hypotheses, the group returned to the engineering design process to plan how they would test and improve their hardware creations so that KIBO could draw (i.e., attaching pens and markers securely to their KIBO, see Fig. 10.10) and how they would create programs that would make KIBO draw the different shapes (i.e., which blocks would make KIBO draw a square versus a circle, see Fig. 10.11). After a few experiments with robots and children's own movements, the group agreed that KIBO could make curves using "turn" movements and its "spin" block (see Fig. 10.12). This moment of math exploration, driven purely by children's curiosity, is reminiscent of the turtle-geometry described by Papert (1980), in which he discusses how children using LOGO who made many small straight lines eventually discovered that by changing angles, they could create different shapes and even curves and circles.

Fig. 10.10 Children programmed their robot to draw a straight line on paper

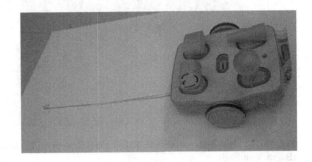

Fig. 10.11 Children combined lines to create shapes such as triangles, stars, and squares with KIBO

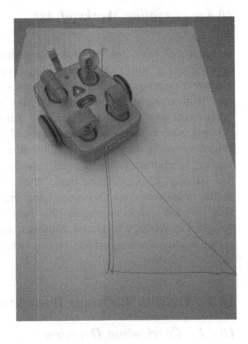

Fig. 10.12 This figure shows the KIBO robot drawing a circle and its accompanying program (Begin Spin End)

10.3.6 Creating Works of Art

Last but not least, the Art-Bots curriculum focused on artistic expression and design (see Fig 10.13). As a group they investigated different artistic styles such as watercolor paintings, abstract art, and photo-realism as expressed by favorite picture book illustrators such as Eric Carle and Lois Ehlert. Children were then given time to freely create any illustrations they chose using KIBO to guide their art materials. In planning how to make their "robot artists," children considered the techniques that illustrators might use to achieve different effects (e.g., short movements and long winding brushstrokes), and the programming blocks necessary to capture the same look (see Fig. 10.14). Children moved their own arms and bodies, and spoke out loud as they considered what they were doing and how they would instruct a machine to carry out these motions. Again, this recalls the self-reflective thinking strategies described by Papert, who argued that these "meta-cognitive" learning opportunities are unique to programming and robotic experiences (Papert 1980).

10.4 Vignette 2: Dances from Around the World

10.4.1 Curriculum Overview

In this vignette, the arts are explored with KIBO through a formal curricular unit called "Dances from Around the World." The Dances from Around the World unit

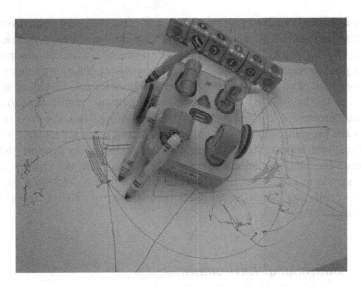

Fig. 10.13 Children created visual works of art using traditional art tools attached to the front and sides of KIBO

Fig. 10.14 This painting was made using paintbrushes taped onto KIBO

is designed to combine music, culture, dance, and language with programming and engineering content. While the Art-Bots activity was completed during one six-hour Saturday session, the Dances from Around the World unit was covered over the course of approximately seven weeks in a formal school setting. Each week, teachers introduced new robotics and programming concepts, from basic sequencing through conditional statements, to their students within the curriculum's music and dance theme. For example, in order to teach the concept of sequencing, children programmed their robots to dance to the Hokey-Pokey with them.

Lessons took place for approximately one hour once a week, leading up to a final project. For the final project, students worked in pairs or small groups to design, build, and program a cultural dance of their choosing. This involved not only robotics and programming knowledge, but also research into the music, the history and cultural relevance of the dance, and facts about the country in which the dance originated. The unit culminated in a dance recital for both the children and the robots to perform in together. Finally, their hard work was celebrated when they received certificates showcasing their knowledge of KIBO robotics and engineering.

10.4.2 Educational Environment

The Dances from Around the World curriculum unit was developed by the DevTech Research Group at Tufts University. It has been used with a variety of schools, camps, and after-school programs across the USA. The unit is now freely available on DevTech's online community the Early Childhood Robotics Network (http://www.tkroboticsnetwork.ning.com) and is therefore available to educators nation and worldwide to adapt and use in their own classrooms. The unit was originally developed for use in an urban public school in Boston, USA, during the piloting phase of the KIBO robot. However, it was recently adapted by five preschool centers in Singapore integrating robotics into their classrooms for the first time. The Singaporean preschools serve as the setting for this vignette.

In order to address the growing need for promoting technological literacy in early childhood classrooms, Singapore's newly launched PlayMaker Programme was released as part of a master plan to introduce younger children to technology (Digital News Asia 2015; Chambers 2015). As part of the PlayMaker Programme initiative, approximately 160 preschool centers across Singapore were given innovative new technological toys that engage children with robotics, programming, building, and engineering including: BeeBot, Circuit Stickers, and KIBO robotics (Chambers 2015). In addition to the new tools, early childhood educators also received training at a one-day symposium on how to use and teach with each of these tools (Chambers 2015).

For the five centers exploring KIBO, the Dances from Around the World unit was chosen because it ties in naturally with the multicultural and bilingual Singaporean community. Singapore has four official languages (English, Mandarin, Tamil, and Malay) and a bilingual education policy where all students in public government schools are taught English as their primary language. However, students also learn another language called their "Mother Tongue," in addition to English. This mother tongue might be Mandarin, Malay, or Tamil depending on the community the school is located in. Because Singaporean children speak different languages and have different cultural backgrounds, the Dances from Around curriculum easily integrated into the cultural appreciation and awareness units already taught in the preschool classes.

10.4.3 STEAM Concepts Addressed

Students covered different STEAM concepts each week, primarily related to technology, engineering, music, and culture. For their final dancing robots, children worked in groups of 2–3 and utilized their existing knowledge of KIBO to create robotic projects that represented the cultural tradition of their choosing.

10.4.4 Music and Dance

Throughout the curriculum children were invited to listen to music drawing on a variety of cultural styles. Varied examples of music and choreographed dances from international traditions, such as the Chinese Lion Dance and Indian Bhangra, were presented to children during play and lesson times. While listening to songs, children spontaneously danced along and became inspired by the videos they had watched with their classmates. This exploration of sound, rhythm, and performance led children to consider dance and elements of movement. They explored behind-the-scenes elements of performance, such as the dancers underneath the Chinese Lion costume.

Later in the curriculum, children became dance choreographers and directors as part of their robotic exploration. Children and teachers chose songs from the class's earlier music explorations and programmed their robots to complete special dance moves in time to the songs. After their robot was programmed the way they wanted, children choreographed their own moves to act out along with the robot. In this way, children explored elements of stage production, as they practiced being live performers, creative directors of the performance, and even the engineers who ensured that the technology and equipment (i.e., the robot and program) were ready for the show.

10.4.5 Repeat Loops and Patterns

While children were free to choose from any of the blocks in KIBO's programming language, many chose to use the Repeat blocks in order to choreograph a dance for their robot that included repetition and patterns. Repeat loops are a foundational concept in computer programming that refers to a sequence of instructions that is continually repeated until a certain condition is reached. In the lessons leading up to the final project, children explored what it means to repeat something and how to do this with KIBO's Repeat blocks. The repeat loops are considered an advanced KIBO programming concept because it requires mastery of a new syntactical rule: KIBO will only repeat what is *between* the "Repeat" and "End Repeat" blocks (See Fig. 10.15). Finally, the Repeat loops must be modified with parameters that dictate how many times the sequence will repeat. These can be numerical parameters or sensor parameters. For example, with a numerical parameter, a child might program KIBO to shake three times. With a sensor parameter, a child might program KIBO to shake until it senses it is near something.

As children worked on their final projects, they drew on their knowledge of Repeat Loops in order to create programs that matched the songs (or clips of songs) their robots would be dancing to. In order to match the song's duration, children experimented with different number parameters that would achieve the correct length of KIBO dance time. Other children used the Repeat blocks to create a dance with only certain dance moves that were repeated, and others that happened just once. The end result was a diverse display of complex robot choreography with a mix of repeated and isolated dance moves.

10.4.6   Expressions of Culture

Children and teachers used the Dances from Around the World curriculum to explore the Chinese, Indian, and Malay cultures that comprise most of the population of Singapore. This manifested itself in different ways for the varying preschool classrooms. For example, some classrooms spent time learning about the foods, clothing, and languages unique to each culture in addition to the music and dance. Children were also encouraged to think about and explore their own cultural backgrounds through discussions with their families at home.

Fig. 10.15 This photograph shows a sample KIBO program using repeat loops. In this program, KIBO would shake 3 times, but only beep once because the Beep block is outside of the Repeat loop

For the final dance recital, students found a variety of ways to express the culture that inspired their dancing robots. Some groups focused on cultural "clothing" for their robot by using arts, crafts, and recycled materials to create performing costumes for KIBO (see Figs. 10.16 and 10.17). Others focused primarily on programming dance moves to accurately resemble the actions of the dance they studied. Still other groups took a more immersive approach to representing the culture they explored with their robotics projects. These students wore clothing to represent the culture of their dance, learned portions of songs themselves, and danced along with their KIBO robots at the recital (see Fig. 10.18).

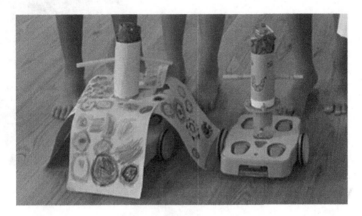

Fig. 10.16 Child-made KIBO costumes for recital

Fig. 10.17 Child-made KIBO puppets for recital

Fig. 10.18 Child wears
traditional Indian clothes
while performing a dance
with KIBO

10.5 Vignette 3: Superhero Bots

10.5.1 Curriculum Overview

From Superman to The Incredibles, children have always had a fascination with
superhero (and super-villain) characters (Jones 2008). The Superhero Bots unit
incorporates an exploration of programming and robotics (with a special emphasis
on sensors) and integrates it with an investigation of superheroes from an interpersonal perspective. This unit leans more on the humanities and civics portion of
the arts, by engaging children in discussions of leadership and decision-making.
Also, since superheroes are typically rooted in back-stories and sagas involving
other characters and important moments in their lives, this unit involves drama and
storytelling elements that enrich the meaning of the robotic creation that children
produce.

10.5.2 Educational Setting

The Superhero Bots curriculum was recently implemented in a one-week robotics
summer camp for children entering kindergarten through second grade. The camp
met for five half-days (approximately four hours each day) and culminated in an

open-house showcase of the children's work that parents, families, friends, and babysitters were invited to attend. Each camp group consisted of approximately 8–10 children and was taught by a college undergraduate or graduate student studying education and technology.

The camp atmosphere provided an informal play and learning environment, which made the superhero content appropriately light-hearted and fun. With the loosely structured days, counselors had time to indulge in extended fantasy and free-play time, which added to the joy of the content. In addition to the time spent on deeper discussions of story structures and personal character, children crafted silly superhero masks and capes, and imagined superhero powers and identities for themselves. This imaginary play helped inspire the robotic constructions that students created for their final projects.

10.5.3   STEAM Concepts Addressed

Children attending this camp were able to spend several hours at a time exploring the robotic components (interspersed with non-robotic games and activities), and they quickly progressed to the more complex elements of the KIBO. Children in the robotics camp explored how to program the sound, distance, and light sensor modules to react to stimuli using conditional "If" statement blocks. In this way, they were able to create interactive robotic creations that could react to the surrounding environment. This exploration of advanced programming blocks was integrated with a discussion on what it means to be a hero and rehearsals for their showcase at the end of the week.

10.5.4   Identity and Civic Engagement

When beginning to work on the superhero projects, the children had a group discussion with their counselor in order to answer this question: What makes someone a hero? Initially children focused on super-abilities such as flying, super strength, and invisibility. However, when prompted to think of some super villains or classic "bad guys" who *also* had many of these super-abilities, the children came to a new conclusion: superheroes strive to "do good" in the world. This child-led discussion naturally came to the conclusion that there are also "everyday superheroes" in the world and they brainstormed a long list that included firefighters, teachers, doctors, and even their parents and friends to inspire the superhero robots they would design for their projects (see Fig. 10.19).

While many superhero movies and shows portray conflicts as black and white, good or bad, the children in this camp came to a conclusion that everyone faces choices each day and that even good people can make mistakes and bad decisions. Finally, they talked about the ways that they can be "everyday heroes" through their

Fig. 10.19 Children's ideas
about what makes a superhero

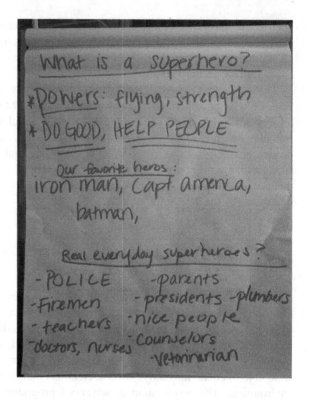

choices in actions such as being a good friend, helping at home, and recycling. These actions were used to help children brainstorm their own superhero characters and personalities that would be brought to life through KIBO.

10.5.5 Sensing Robots

The sensor components of the KIBO robot are among the most complex and engaging parts of the KIBO kit. To introduce the concept of a sensor, counselors first discussed human body parts and our own five senses that allow us to take in information about our environments. Similarly, KIBO's distance, light, and sound sensors allow the robot to take in information about the environment.

Because the children had already expressed their fascination with their favorite superheroes' powers, the sensors were introduced as KIBO's "super senses" that allow it to perform extraordinary tasks. As children were designing their Super Bots to react to the environment in order to help others, they took into consideration the skills that KIBO's different sensors provide. For example, one boy used the KIBO's ear-shaped sound sensor and the accompanying Wait For Clap block. This, he explained in his accompanying story, was how his robot hero "listens for calls for

Fig. 10.20 Robot with light sensor and light output, and program to turn on lightbulb in the dark

help." Another child used the light sensor to detect when it was dark out. When KIBO sensed it was dark, it was programmed to turn on a helpful bright light (using KIBO's light output) to patrol for "bad guys" and guide others to safety (see Fig. 10.20). By using the superhero context which was engaging to the campers, children were motivated to use these complex robotic and programming elements in order to continually improve and refine the design of their robots.

10.5.6 Storytelling and Drama

With their knowledge of KIBO's special sensing abilities as well as a discussion about civic engagement and what it means to be a hero, children let their imaginations run wild as they planned out their own KIBO robots. This process involved discussion and brainstorming, drawing characters, and planning out the KIBO programming blocks that would be used to bring their superhero robots to life. They used arts, crafts, and recycled materials to build their final superhero robots and spent several hours working to program their robot in a way that would showcase its special abilities (see Fig. 10.21).

During this process, the counselors also encouraged children to engage in dramatic free-play that involved making capes, dressing up as superheroes, and constructing an imaginary world populated their favorite heroes. They were also encouraged to think of a story for their super KIBOs, and they read popular children's books around the superhero theme for inspiration. Children were prompted

Fig. 10.21 Example of
KIBO superhero design

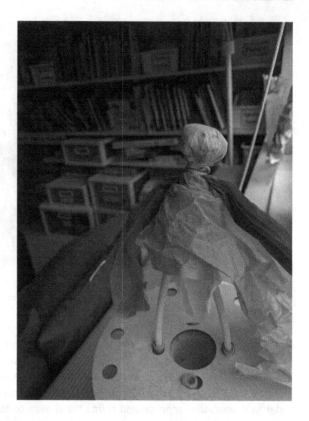

to come up with a beginning, middle, and end of their superhero robot stories and use their program to represent one scene from their story. Along with the programs, children wrote, narrated to a counselor, or drew their completed stories to share with one another. Children came up with both individual and full-group superhero tales.

Throughout the week of the camp, there was also an air of excitement as campers knew that on the last day they would be performing their superhero stories, showcasing their robots, and sharing their knowledge with a crowd of guests made up of their families and friends. This provided an external motivation to complete their robots and stories to the best of their abilities. It also provided an opportunity to take their superhero stories from a written and/or illustrated format, and expand it with some performance elements. The camp groups held several rehearsals to practice songs they had made up, share the superhero stories they invented as a class, and give demonstrations of their robots. The campers focused on projecting their voices, conveying the emotion and personalities of the characters they invented, and properly demonstrating the technical elements of KIBO. In the end, the showcase served as a joyful celebration of the kids' hard work and dedication to their projects.

10.6 Discussion

10.6.1 Strengthening STEM Curriculum Through the Arts

The core element present throughout these three vignettes was the use of the arts to strengthen the STEM learning and exploration that was already happening with KIBO. For example, in the Art-Bots unit, children explored visual art by programming robots to create drawings. In this case, working toward their artistic goal (be it a shape, letter, or abstract picture) prompted them to explore mathematical concepts such as geometric paths, straight and curved lines, and angles. This also prompted a deep exploration of sturdy building and iterative engineering design.

Research has shown that music and movement can be beneficial to young children's development (Andress 1980; Lillard 2005). The Dances from the Around World unit invited children to investigate movement and music from different cultures. Although children themselves might not be able to act out all of the ritualistic dances, they were able to break apart complex steps into smaller parts that even a KIBO robot could do. In this way, children exercised their sequencing skills, a foundational skill for both developing numeracy and literacy as well as computer programming.

Finally, storytelling and drama were used to strengthen the KIBO unit on superheroes. In this camp, dramatic play offered children the chance to use symbolic representation through engaging in make-believe play with familiar objects such as blankets representing capes, and legos representing cities to protect. Dramatic play can also prompt literacy development when it involves the use of reading and writing materials (Christie 1990; Fields and Hillstead 1990). In the superhero robots camp, children explored books and stories with their counselors, created lists and brainstorms, and developed their own sequential stories using a combination of writing, drawing, and speaking. The stories provided a context for the programs the children made for their superhero KIBOs, and prompted them to think logically and sequentially when constructing these programs. Throughout this process of programming and storytelling, children were immersed in a world of fantasy and role-playing that captured their attention for intensive learning in the form of light-hearted play.

10.6.2 The Educational Environment

The three vignettes in this chapter highlighted a range of different types of teaching and learning environments. In the Art-Bots curriculum, we saw educational technology specialists leading longer workshop-style explorations for a mixed-age group of children. The Dances from Around the World curriculum showed formal early childhood educators teaching sequential lessons in their classrooms each week over the course of nearly two months. Finally, in the Superhero Bots curriculum we saw college student counselors leading a play-based exploration of robotics and the dramatic arts. Each of these environments offered different strengths to the STEAM content being taught. In the STEM-Saturday club, children played at being "scientists" engaged in secret experiments, with volume and buoyancy, map-making, and bridge construction being some of the many exciting mysteries they explored along with KIBO robotics. The Dances unit, which lasted almost two months, gave children the space to bring cultural inspiration from home to inform their song and dance performances. The Superhero Bots curriculum offered its own playful approach, with counselors and kids using robots as part of a week-long imaginative play experience.

When designing an educational environment for young children, it is important to consider a few questions. First, what are the educational needs of the students? Before designing any kind of educational experience, especially one that involves technology, it is important to identify your learning goals and determine how technology and the arts will serve to *enhance* children's learning rather than *distract* from it. In the case of the three vignettes in this chapter, the use of KIBO robotics served as a technical medium to teach foundational engineering and programming content. But it also served to help bring big ideas about culture, identity, and the visual arts to life in a tangible and interactive way that held the attention of the kids.

Secondly, educators will want to consider the resources available to them. Using the KIBO robotics kits as an example, if a classroom only has 1–2 robotics kits, setting up a robotics center or a technology corner might be the most effective environment for learning. Children in the Dances from Around the World curriculum worked effectively in groups of 2–3 children. While this did spark some of the usual conflicts and discomforts that characterize group work with preschool aged children, it also provided an opportunity for practicing collaboration, communication, and troubleshooting conflict in an authentic way.

Finally, educators should consider the physical space where the children will engage with the new technology. Teachers leading the Dances curriculum used different spaces to achieve different ends, giving children small, isolated work areas to become acquainted with the robot kit, and bringing them to large open areas where they could move freely when they were choreographing and practicing their robot dances. In any environment, it is important to consider ways to help the children feel ownership and safety, by posting their hand-drawn robot designs, or displaying robotic and art objects to further inspire children's designs.

10.6.3   The Teacher's Role

These vignettes predominantly highlighted the children's experiences and the products they made throughout the three curriculum units. However, it is important to highlight the adult's role in the classroom when working with technology and arts. Each of these units took a *constructionist* approach to teaching and learning.

Constructionism is the idea that people learn effectively through making things (Papert 1980; Ackermann 2001). When considering that children learn through making tangible objects as well as testing their own theories and ideas, this puts teachers in a special role. In these types of learning environments, it is often useful to think of the teacher as more of a facilitator than an instructor. For example, in the Superhero Bots camp, the counselors followed the discussions and interests of the children and used these conversations to help guide the expectations for their final robotics projects. Similarly, when it came to providing technical support to children grappling with difficult concepts such as sensors and repeat loops, teachers from all three units worked to provide prompts, examples, and demonstrations rather than step-by-step answers and tutorials. This type of facilitation allowed the children to have their own "aha!" moments while making discoveries about the technology.

This Constructionist approach to teaching and learning is aligned with Bers (2012) Positive Technological Development (PTD) framework, which guided the development of each of the curricular units presented in this chapter. Through the PTD approach, children not only gained skills related to technology, math, and the arts, but they also gained interpersonal skills from working in groups and sharing materials. Children learned how to effectively share their thoughts and ideas with the greater community. For example, in both the Dances and Superhero curricula, the units culminated with a showcase that was open to friends, parents, and family members to learn about KIBO from the children. In this way, robotics and the arts were in service of greater early childhood developmental needs such as community building and fostering a caring and supportive environment for playful inquiry.

10.7 Conclusion

There is often a misconception that the end goal of science and technology curricular interventions is to prepare children to grow up and become scientists, mathematicians, and engineers. Quite the opposite, the goal of the PTD approach to STEAM education is to provide young children with a mindset that is applicable to a range of subject matter and experiences they have during their schooling years and beyond (Bers 2012). While STEM career fields are rapidly growing in the USA, in future decades, many of our best leaders may come from art and design backgrounds (Maeda 2012). Whether today's kindergarteners grow up to become ballerinas, inventors, designers, or teachers, their success will be rooted in an ability to problem-solve and think creatively. By integrating the arts with technical and scientific fields starting in early childhood, young children grow up with the abilities they need to be well-rounded thinkers in any domain they pursue.

References

Andress, B. (1980). *Music experiences in early childhood*. New York, NY: Schirmer Books.

Ackermann, E. (2001). Piaget's constructivism, Papert's constructionism: What's the difference. *Future of Learning Group Publication, 5*(3), 438.

Althouse, R., Johnson, M. H., & Mitchell, S. T. (2003). *The colors of learning: Integrating the visual arts into the early childhood curriculum*. New York: Teachers College Press.

American Academy of Pediatrics. (2003). Prevention of pediatric overweight and obesity: Policy statement. *Pediatrics, 112*, 424–430.

Bers, M. U. (2008). *Blocks to robots: Learning with technology in the early childhood classroom*. NY: Teachers College Press.

Bers, M. U. (2010). The TangibleK Robotics Program: Applied computational thinking for young children. *Early Childhood Research and Practice, 12*(2).

Bers, M. U. (2012). *Designing digital experiences for positive youth development: From playpen to playground*. Oxford: Oxford University Press.

Bers, M. U. (2017). *Coding as a playground: Programming and computational thinking in the early childhood classroom*. Routledge press.

Bers, M. U., Ponte, I., Juelich, K., Viera, A., & Schenker, J. (2002). Teachers as designers: Integrating robotics into early childhood education. *Information Technology in Childhood Education*, 123–145.

Bers, M. U., Seddighin, S., & Sullivan, A. (2013). Ready for robotics: Bringing together the T and E of STEM in early childhood teacher education. *Journal of Technology and Teacher Education, 21*(3), 355–377.

Brosterman, N. (1997). *Inventing kindergarten*. New York: H.N. Abrams.

Cejka, E., Rogers, C., & Portsmore, M. (2006). Kindergarten robotics: Using robotics to motivate math, science, and engineering literacy in elementary school. *International Journal of Engineering Education, 22*(4), 711–722.

Chambers, J. (2015). Inside Singapore's plans for robots in pre-schools. *GovInsider*.

Christie, J. F. (1990). Dramatic play: A context for meaningful engagements. *The Reading Teacher, 43*(8), 542–545.

Clements, D. H. (1999). Young children and technology. In G. D. Nelson (Ed.), *Dialogue on early childhood science, mathematics, and technology education*. Washington, DC: American Association for the Advancement of Science.

Cordes, C., & Miller, E. (2000). *Fool's gold: A critical look at computers in childhood*. College Park, MD: Alliance for Childhood.

Digital News Asia. (2015, September 24). IDA launches S$1.5 m pilot to roll out tech toys for preschoolers. Retrieved from: https://www.digitalnewsasia.com/digital-economy/ida-launches-pilot-to-roll-out-tech-toys-for-preschoolers.

Fields, M. V., & Hillstead, D. V. (1990). Whole language in the play store. *Childhood Education, 67*(2), 73–76.

Hobbs, R. (2010). Digital and media literacy: A plan of action. The Aspen Institute.

Hollandsworth, R., Dowdy, L., & Donovan, J. (2011). Digital citizenship in K-12: It takes a village. *TechTrends, 55*(4), 37–47.

Jones, G. (2008). *Killing Monsters: Why children need fantasy, superheroes, and make-believe violence*. Basic Books.

Kazakoff, E., & Bers, M. (2012). Programming in a robotics context in the kindergarten classroom: The impact on sequencing skills. *Journal of Educational Multimedia and Hypermedia, 21*(4), 371–391.

Kuh, L. P. (Ed.). (2014). *Thinking critically about environments for young children: Bridging theory and practice*. New York, NY: Teachers College Press.

Lee, K., Sullivan, A., & Bers, M. U. (2013). Collaboration by design: Using robotics to foster social interaction in Kindergarten. *Computers in the Schools, 30*(3), 271–281.

Lerner, R. M., Lerner, J. V., Almerigi, J., Theokas, C., Phelps, E., Gestsdottir, S., et al. (2005). Positive youth development, participation in community youth development programs, and community contributions of fifth grade adolescents: Findings from the first wave of the 4-H study of positive youth development. *Journal of Early Adolescence, 25*(1), 17–71.

Lillard, A. (2005). The impact of movement on learning and cognition. In A. Lillard (Ed.), *Montessori: The science behind the genius*. New York, NY: Oxford University Press.

Maeda, J. (2012). *STEM to STEAM: Art in K-12 is key to building a strong economy*. Edutopia: What works. in education.

Maguth, B. (2012). In defense of the social studies: Social studies programs in STEM education. *Social Studies Research and Practice, 7*(2), 84.

Massachusetts Department of Education. (2006). Massachusetts science and technology/engineering curriculum framework. Retrieved from http://www.doe.mass.edu/frameworks/scitech/1006.pdf

Nicholson, S. (1972). The theory of loose parts, an important principle for design methodology. *Studies in Design Education Craft & Technology, 4*(2).

Oppenheimer, T. (2003). *The flickering mind: Saving education from the false promise of technology*. New York: Random House.

Papert, S. (1980). Mindstorms: Children, computers, and powerful ideas. Basic Books, Inc.

Perlman, R. (1976). Using computer technology to provide a creative learning environment for preschool children. Logo memo no 24, Cambridge, MA: MIT Artificial Intelligence Laboratory Publications 260.

Pomeroy, S. R. (2012). From STEM to STEAM: Science and art go hand-in-hand. *Scientific American Guest Blog*.

Resnick, M. (2006). Computer as paintbrush: Technology, play, and the creative society. *Play = learning: How play motivates and enhances children's cognitive and social-emotional growth*, 192–208.

Resnick, M., Martin, F., Berg, R., Borovoy, R., Colella, V., Kramer, K., et al. (1998). Digital manipulatives. In *Proceedings of the CHI '98 Conference*, Los Angeles, April 1998.

Robelen, E. W. (2011). STEAM: Experts make case for adding arts to STEM. *Education week, 31* (13), 8.

Root-Bernstein, R. (2011). The art of scientific and technological innovations. *Retrieved* April, 13, 2011.

STEM to STEAM. (2016). Retrieved July 27, 2016, from http://stemtosteam.org/

Strong-Wilson, T., & Ellis, J. (2007). Children and place: Reggio Emilia's environment as third teacher. *Theory Into Practice, 46*, 40–47.

Sullivan, A., & Bers, M. U. (2017). Dancing robots: Integrating art, music, and robotics in Singapore's early childhood centers. *International Journal of Technology and Design Education*. Online First.

Sullivan, A., & Bers, M. U. (2015). Robotics in the early childhood classroom: Learning outcomes from an 8-week robotics curriculum in pre-kindergarten through second grade. *International Journal of Technology and Design Education*. Online First.

Sullivan, A., Elkin, M., & Bers, M. U. (2015). KIBO Robot demo: Engaging young children in programming and engineering. In *Proceedings of the 14th International Conference on Interaction Design and Children (IDC '15)*. ACM, Boston, MA, USA.

Sullivan, A., Kazakoff, E. R., & Bers, M. U. (2013). The wheels on the bot go round and round: Robotics curriculum in pre-kindergarten. *Journal of Information Technology Education: Innovations in Practice, 12*, 203–219.

U.K. Department for Education. (2013, September). *National curriculum in England: Computing programmes of study*. Statutory guidance. London, UK: Crown copyright.

U.S. Department of Education, Office of Educational Technology (2010). *Transforming American education: Learning powered by technology*. Washington, D.C. Retrieved from http://www.ed.gov/technology/netp-2010

White House. (2011). *Educate to innovate*. Retrieved from: http://www.whitehouse.gov/issues/education/educate-innovate

Wing, J. (2006). Computational thinking. *Communications of the ACM, 49*(3), 33–35.

Wyeth, P. (2008). How young children learn to program with sensor, action, and logic blocks. *International Journal of the Learning Sciences, 17*(4), 517–550.

Yakman, G. (2008). STEAM education: An overview of creating a model of integrative education. In *Pupils' Attitudes Towards Technology (PATT-19) Conference: Research on Technology, Innovation, Design & Engineering Teaching*, Salt Lake City, Utah, USA.

Index

A
Arts, 231, 232, 234, 238, 244, 250, 253, 255–257
Authentic learning, 63, 64
Automation, 171, 172, 174–178, 180, 181, 183, 187–192

C
Coding, 131, 132, 167
Computer programming, 59–63, 65, 74, 75
Computing, 131, 132, 146
Concepts, 103, 107, 109, 113, 115–117, 120, 124
Constructionism, 10, 11, 13
Creativity, 213–215, 218, 219, 221, 222, 224–228
Curriculum, 34, 35, 39, 43, 44, 50–54

D
Design, 213, 214, 216, 218–221, 225–227
3D printing, 92

E
Early childhood, 232–235, 237, 239, 246, 255, 257
Education, 195, 197, 198, 200, 202, 205, 207
Educational robotics, 3, 8–12, 14–18, 20–23, 25, 26

H
Humanities, 231, 234, 250

I
Innovation literacy, 18
Innovative tools, 59, 69
Interface, 94, 98

L
Learner-centred learning, 65
Learning, 195, 197, 198, 201, 203, 205
Lego, 196, 198, 200
Lego Mindstorms robotics, 67, 69
Lego robots, 120, 125

M
Maker movement in education, 12, 14, 15
Methodology, 109, 113, 125

O
Operating system, 91

P
Play, 213, 214, 216–218, 227, 228
Problem-Based-Learning (PBL), 132, 134, 168
Problem-solving, 213, 214, 219, 221, 227
Programming, 113, 115, 117, 118, 120

R
Review, 103–109, 111, 114, 117, 118, 123, 124
Robotics, 33–37, 39, 40, 42–45, 49, 50, 52–54, 85–87, 89–92, 98–100, 132, 133, 139, 144, 147, 155, 161–164, 167, 168, 171–183, 186, 187, 189–192, 195–198, 200, 205, 207, 213, 216–219, 221–228, 231–235, 239, 246, 250, 256, 257

S
Simulation software, 173, 180, 186, 189, 192
STEAM, 231–234, 236–239, 247, 251, 256, 257
STEM+C, 131, 132, 155, 159, 161, 162, 165, 167, 168

© Springer International Publishing AG 2017
M.S. Khine (ed.), *Robotics in STEM Education*,
DOI 10.1007/978-3-319-57786-9

STEM, 33–44, 49–51, 53, 54, 90, 100,
     195–198, 201, 205, 207
STEM education, 171, 174, 186
Systems thinking, 34, 35

T
Technological literacy, 17, 19
Technology, 103–107, 115, 117, 121, 122,
     124, 195–197

Printed in the United States
By Bookmasters